AUTHOR	CLASS
BRISTOW, C.R.	554.2238

TITLE Geology of the country around
Royal Tunbridge Wells.

PLATE I (*Frontispiece*)

SCENERY ASSOCIATED WITH THE OUTCROP OF THE ASHDOWN BEDS AT COLEMAN'S HATCH (A9912)

NATURAL ENVIRONMENT RESEARCH COUNCIL

INSTITUTE OF GEOLOGICAL SCIENCES

MEMOIRS OF THE GEOLOGICAL SURVEY OF GREAT BRITAIN

ENGLAND AND WALES

Geology of the Country around Royal Tunbridge Wells

(Explanation of One-inch Geological Sheet 303, New Series)

by

C. R. Bristow, Ph.D. and R. A. Bazley, Ph.D.

with contributions by

R. W. Gallois, B.Sc., E. R. Shephard-Thorn, Ph.D.;
R. Allender, Ph.D. (Water Supply);
R. Casey, Ph.D., F.R.S. and
H. C. Ivimey-Cook, Ph.D. (Palaeontology)

LONDON
HER MAJESTY'S STATIONERY OFFICE
1972

PREFACE

THIS MEMOIR describes the geology of the country depicted on the Tunbridge Wells (303) Sheet of the One-inch Geological Map of England and Wales. The district was first geologically surveyed on the one-inch scale by W. T. Aveline, H. W. Bristow, W. B. Dawkins, F. Drew, C. Gould, C. Le Neve Foster and W. Topley and the results published in 1864 on Old Series One-inch Sheets 5 and 6. A revised edition of Sheet 5 was published in 1893 and revised editions of Sheet 6 were published in 1886 and 1889. Topley's classic memoir 'Geology of the Weald' which describes the geology of Sheets 5 and 6 and adjacent sheets, was published in 1875.

Several of the Geological Survey 'Water Supply' memoirs have dealt in part with the district, namely: 'Water Supply of Sussex' by W. Whitaker and Clement Reid (1899) and a supplement to it by Whitaker (1911), 'Water Supply of Kent' by Whitaker (1908), 'Water Supply of Surrey' by Whitaker (1912) and 'Wells and Springs of Sussex' by F. H. Edmunds (1928). Wartime Pamphlet No. 10, Part VII, compiled by Messrs. S. Buchan, J. A. Robbie, S. C. A. Holmes, J. R. Earp, E. F. Bunt and L. S. O. Morris, catalogues the wells of the Tunbridge Wells sheet and was published in 1940. A revised well catalogue, covering sheets 303 and 304 (Tenterden), prepared by Miss B. I. Harvey and Miss A. M. Matthews, was published in 1964.

The primary six-inch geological survey of the Tunbridge Wells area was made by Messrs. C. R. Bristow, R. A. Bazley, E. R. Shephard-Thorn, R. W. Gallois, G. Bisson and E. A. Edmonds between 1960 and 1965, under the direction of Mr. S. C. A. Holmes as District Geologist. A small part of the Tunbridge Wells area was surveyed on the six-inch scale by Dr. S. Buchan in 1933–36 and Dr. F. B. A. Welch in 1931 as an overlap from the Sevenoaks (287) Sheet; this ground has been resurveyed in the recent investigation by Dr. C. R. Bristow. Lists of six-inch maps and the names of the surveyors are given on pp. ix, x. The one-inch geological map of the area was published in 1971. The memoir has been compiled by Dr. Bristow from the accounts of the surveyors, with contributions by Dr. R. Casey and Dr. H. C. Ivimey-Cook on the palaeontology of the Ashdown boreholes and by Dr. R. Allender on the water resources. It has been edited by Dr. Bristow and Mr. Holmes.

We are grateful for the assistance of Dr. J. Callomon and Dr. W. G. Chaloner for their identifications of ammonites and spores from the Ashdown boreholes. The remaining fossils from these boreholes have been identified by members of the Palaeontology Department, namely Dr. Ivimey-Cook, Dr. F. W. Anderson and Dr. R. Casey.

We wish to thank BP Petroleum Development Limited for permission to examine the cores and to publish details of their Ashdown Nos. 1 and 2 boreholes.

Thanks also are given to Messrs. Herbert Lapworth Partners for permission to make use of their geological section across the Weir Wood Reservoir site (Fig. 6) and to the North West Sussex Water Board for permission to publish this section; also to the Borough of Royal Tunbridge Wells for permission to examine cores and publish details of boreholes in their area.

The surveyors are grateful for the ready co-operation which they received from the landowners and tenants of the district in facilitating access to their properties.

K. C. DUNHAM
Director

Institute of Geological Sciences
Exhibition Road
South Kensington
London, SW7
3rd February, 1972

CONTENTS

ILLUSTRATIONS

TEXT-FIGURES

EXPLANATION OF PLATES

[1]Numbers preceded by A refer to photographs in the Geological Survey collection.

LIST OF SIX-INCH MAPS

The following is a list of six-inch geological maps included in the area of One-inch Geological Sheet 303, with the initials of the surveying officers and the date of the survey for each six-inch map; the officers are: R. A. Bazley, G. Bisson, C. R. Bristow, E. A. Edmonds, R. W. Gallois and E. R. Shephard-Thorn.

Manuscript copies of these maps will be deposited for public reference in the library of the Institute of Geological Sciences. They contain more detail than appears on the one-inch map.

TQ 32 NE (part of)	Horsted Keynes	R.W.G.	1961
TQ 32 SE (part of)	Scaynes Hill	R.W.G.	1964
TQ 33 NE (part of)	East Grinstead..	R.W.G.	1962
TQ 33 SE (part of)	West Hoathly	R.W.G.	1961
TQ 34 SE (part of)	Dormans Park	R.W.G.	1962
TQ 42 NW	Danehill	R.W.G., C.R.B.	1961, 1964
TQ 42 NE	High Hurstwood	C.R.B.	1965
TQ 42 SW (part of)	Fletching	R.W.G., C.R.B.	1964
TQ 42 SE (part of)	Maresfield	C.R.B.	1964
TQ 43 NW	Ashurstwood	R.W.G., C.R.B.	1962–63
TQ 43 NE	Hartfield	C.R.B.	1963
TQ 43 SW	Forest Row	R.W.G., C.R.B.	1961, 1963
TQ 43 SE	Coleman's Hatch	C.R.B.	1963
TQ 44 SW (part of)	Blockfield	R.W.G., C.R.B.	1962–63
TQ 44 SE (part of)	Cowden	C.R.B.	1963
TQ 52 NW	Jarvis Brook	R.A.B.	1964
TQ 52 NE	Mayfield	R.A.B.	1963
TQ 52 SW (part of)	Hadlow Down	R.A.B.	1964
TQ 52 SE (part of)	Heathfield	R.A.B.	1964
TQ 53 NW	Groombridge	C.R.B.	1963
TQ 53 NE	Tunbridge Wells	C.R.B.	1963–64
TQ 53 SW	Crowborough	C.R.B.	1963
TQ 53 SE	Mark Cross	R.A.B.	1964
TQ 54 SW (part of)	Fordcombe	C.R.B.	1963
TQ 54 SE (part of)	Speldhurst	C.R.B.	1964
TQ 62 NW	Wadhurst Park	R.A.B.	1963
TQ 62 NE (part of)	Stonegate	R.W.G., R.A.B.	1960, 1963
TQ 62 SW (part of)	Burwash Common	R.A.B.	1963
TQ 62 SE (part of)	Brightling	E.A.E., R.A.B.	1960, 1963–64
TQ 63 NW	Bell's Yew Green	C.R.B.	1964

TQ 63 NE					
(part of)	Lamberhurst	..	..	..	G.B., E.R.S.-T.	1960, 1963
TQ 63 SW	Wadhurst	..	..	..	R.A.B.	1963–64
TQ 63 SE					
(part of)	Ticehurst	..	..	..	E.R.S.-T.	1960, 1963–64
TQ 64 SW					
(part of)	Pembury	..	..	..	C.R.B.	1964
TQ 64 SE					
(part of)	Brenchley	..	..	..	G.B., C.R.B.	1960, 1965

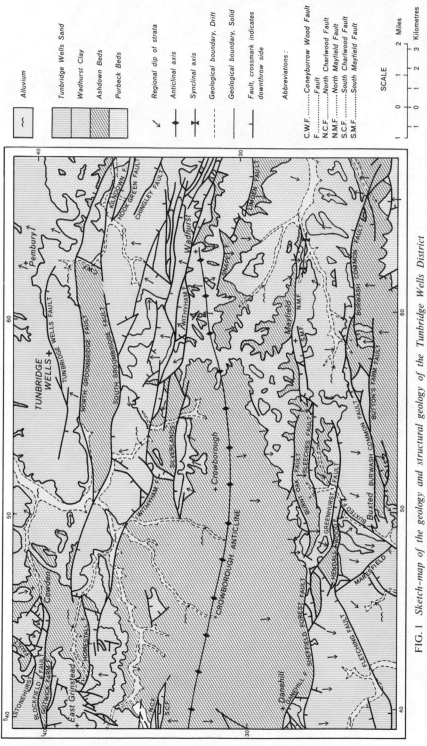

FIG. 1 *Sketch-map of the geology and structural geology of the Tunbridge Wells District*

Chapter I

INTRODUCTION

THE TUNBRIDGE WELLS district is situated in the centre of the High Weald and straddles the county boundaries of Sussex, Surrey and Kent, with the greater part lying in East Sussex. Geologically this area is mostly underlain by Hastings Beds, an alternating sequence of sandstones, silts, clays and limestones (Fig. 1). The clays have their largest outcrop in the south-east corner of the sheet and give rise to heavy, badly drained, clay soils. Elsewhere the soils are light and support only poor farmland, or heath and lightly wooded ground of which the best known tract is Ashdown Forest, a surviving part of the original Wealden Forest, Andredsweald. The Forest roughly coincides with the Ashdown Beds outcrop of the Crowborough Anticline, and is the area from which this formation takes its name.

The community is basically agricultural and the settlements scattered. The only towns of any size within the district (see Fig. 1) are: the historic spa of Royal Tunbridge Wells (pop. 44 000), East Grinstead (pop. 17 000) and the sprawling town of Crowborough (pop. 8000). The two former were originally market centres but all three have now become dormitory towns.

This district was once the leading iron-producing region of Britain. Although iron was first worked in prehistoric times and later by the Romans there was a decline in its use until the late 15th century. From this date the iron industry rapidly increased in importance to reach its peak in the 16th century and then to decline and finally become extinct at the beginning of the 19th century. At the present day there is little evidence of the forges and furnaces, with the exception of the hammerponds and the place names, e.g. Furnace Farm [453 401][1], Forge Farm [533 353] and Cinderhill [603 284].

The major watershed of the High Weald crosses the district from east to west. To the north lies the catchment area of the River Medway draining to the North Sea, while to the south the rivers Ouse and Rother, which flow to the English Channel, drain the south-western and south-eastern portions of the district respectively (see Figs. 2, 6). The relief varies from about 50 ft (15·2 m) O.D. in the south-west of the map to 792 ft (241·4 m) O.D. at Crowborough Beacon, the highest point of the High Weald on the Medway–Ouse watershed. Deeply entrenched and well-wooded valleys draining from the watershed characterize this largely unspoilt area of the High Weald.

GEOLOGICAL SEQUENCE

The formations represented on the one-inch geological map and sections (Sheet 303) are listed below:

---

[1]National Grid references are given in this form throughout. All lie within the 100-km grid square TQ or 51.

1

SUPERFICIAL FORMATIONS (DRIFT)

RECENT AND PLEISTOCENE

    Landslips
    Alluvium
    Terraced River Gravels
    Head

SOLID FORMATIONS

	Generalized thickness	
	Feet	(metres)
CRETACEOUS		
WEALDEN		
HASTINGS BEDS		
Upper Tunbridge Wells Sand: sands, sandstones, silts and impersistent beds of clay .. .. up to	220	(67·1)
Upper Grinstead Clay: red and grey clays with thin siltstones	0–30	(0–9·1)
Cuckfield Stone: sandstones, calcareous sandstones and siltstones	0–30	(0–9·1)
Lower Grinstead Clay: grey clay, with thin limestones, thin calcareous sandstones, ironstone nodules ..	0–30	(0–9·1)
Ardingly Sandstone: massive and thickly bedded sandstone and sandrock	0–70	(0–21·3)
Lower Tunbridge Wells Sand: silts and sandstone; locally thin beds of clay	50–150	(15·2–45·7)
Wadhurst Clay: soft grey mudstone with thin siltstones and silty mudstones, limestones and ironstone seams	110–235	(33·5–71·6)
Ashdown Beds: silts and sands with thin clay seams becoming more important towards the base ..	650–750	(198·1–228·6)
JURASSIC		
Upper Purbeck (including the Greys Limestones): mudstones and shales with shelly limestones ..	106–136	(32·3–41·5)
[1]Middle Purbeck (including the Cinder Beds): mudstones and shales with sandstones and limestones	180–240	(54·9–73·2)
Lower Purbeck (including the Blues Limestones): finely crystalline limestones, calcareous mudstones and gypsiferous strata	195–244	(59·4–74·4)

The following formations have been proved in the Ashdown boreholes:

	Generalized thickness	
	Feet	(metres)
JURASSIC		
Purbeck Beds	620	(189)
Portland Beds	70–80	(21·3–24·4)
Kimmeridge Clay	1730–1839	(527·3–560·5)
Corallian Beds	368–384	(112·2–117·0)
Oxford Clay	304–317	(92·6–96·6)
Kellaways Beds	30	(9·1)
Cornbrash and Forest Marble	33–49	(10·1–14·9)

[1]On faunal grounds the boundary between the Cretaceous and Jurassic Systems has been advocated at the base of the Cinder Beds (Casey 1963).

						Generalized thickness	
						Feet	(metres)
Great Oolite Limestone..	..	..	..	..	..	143–158	(43·6–48·2)
Fuller's Earth	..	..	..	..	..	84–90	(25·6–27·4)
Inferior Oolite	..	..	..	..	..	379–411	(115·5–125·3)
Lias, undifferentiated	..	..	..	..	..	1292	(393·8)

unconformity

Beds of unknown age	..	..	..	..	..	81	(24·7) proved

GEOLOGICAL HISTORY

In this area the Mesozoic rocks were deposited on a surface of folded Devonian, Carboniferous and possibly of New Red Sandstone rocks. A brief history of deposition of the Jurassic strata and a figure showing variations in thickness of the various horizons is given in the Tenterden (304) Sheet Memoir (Shephard-Thorn and others 1966).

The oldest strata that crop out on Sheet 303 are the Purbeck Beds. These were deposited in lagoonal areas probably produced by differential uplift at the close of Portlandian times. The lagoon (Anderson and others 1967, p. 174) had a restricted circulation of water; evaporites (mainly gypsum and anhydrite) were amongst the earliest deposits, and their presence suggests that the climate was hot and dry. Later, connexion with the open sea to the south became stronger, and marine and quasi-marine episodes alternated with brackish and probably fresh-water conditions. The evidence for episodes that were relatively marine comes mainly from the varying ostracod faunas, although foraminifera and the presence of oysters within the Cinder Beds are important. Throughout Purbeck times there was rhythmic sedimentation of limestones, mudstones, silts and sandstones.

The Purbeck Beds are succeeded by sediments having features comparable to those of modern deltas. Within the Ashdown Beds sand gradually became the dominant sediment. The probable source of this non-marine sediment and the related palaeogeography of the period have been discussed in detail by Allen (1967). The Ashdown Beds were succeeded by the predominantly argillaceous sediments of the Wadhurst Clay and these are regarded as having been deposited in a pro-delta shelf environment (Taylor 1963). As in the Purbeck period, salinity fluctuated during this time (Anderson and others 1967).

These conditions were followed by a readvance of the delta front bringing in much sand to form the deposits known as the Tunbridge Wells Sand, the youngest solid formation in the present map area. This includes the Grinstead Clay in the west, and elsewhere occasional thin clay seams. From the surrounding areas it is deduced that Weald Clay, Lower Greensand and Upper Cretaceous sediments were originally deposited all over the area prior to minor earth movements in late Cretaceous times which caused the land to be tilted and subjected to sub-aerial denudation (the overlying Eocene strata have a slightly discordant relationship with the Upper Chalk). Later, more marked orogenic movements took place during the Oligocene–Miocene interval with the formation of the Wealden anticlinorium and associated fault pattern. This resulted in the raising of much of south-east England above sea-level and its subjection to sub-aerial erosion, culminating in the formation of the Mio-Pliocene peneplain at about 800 ft (243·8 m) O.D. During the Pliocene period parts of the

Weald were again submerged, but only the lower ground in the south-west corner of the present area might have been affected, while elsewhere sub-aerial denudation continued. With uplift of the area and consequent relative fall of the Pliocene sea-level primitive rivers extended across, and eventually became established on, the former wave-cut platform (Wooldridge and Linton 1955). During the Pleistocene period periglacial conditions must also have played an important part in rock destruction and transport, although much of the resulting debris has been removed by present-day erosion.

History of Research

The first comprehensive account of the geology of this district was by Topley (1875) in 'Geology of the Weald'. In this he described the main stratigraphical horizons that were first named by Drew (1861) and which are still used today. He also reviewed all the important earlier literature and his detailed work has become a foundation stone on which all later work has been built.

The most recent stratigraphical, structural and palaeontological accounts of all the Purbeck inliers of the Weald are those by Howitt (1964) and by Anderson and Bazley (1971). Summaries of earlier researches are given in both papers. Detailed examination of the Ashdown Beds, which reach 750 ft (228·6 m) in thickness, has so far been limited to the Top Ashdown Pebble Bed and the Top Ashdown Sandstone (Allen 1949a; 1959; 1960b; 1961).

It is only recently that a full sequence of the Wadhurst Clay has been investigated, following trial boreholes near Wadhurst (Anderson and others 1967). Attention in the past has, of necessity, centred on the isolated exposures with soil beds, bone beds and their associated strata (Allen 1941; 1947). Milner (1923a, b) used sedimentary analysis as a means of correlating the Hastings Beds, and in particular the Tunbridge Wells Sand. Later work by Buchan (1938) and Bazley and Bristow (1965) has, however, shown the necessity for revising his stratigraphical correlations. Origin of the 'sandrock' tors (Ardingly Sandstone of Gallois 1965a) has been discussed by Bird (1964). The provenance and petrology of the Top Lower Tunbridge Wells Pebble Bed have been examined by Allen, whose attempts to reconstruct the palaeogeography of the Hastings Beds (Allen 1954; 1959; 1960b; 1961; 1967b) have been supported by Taylor (1963) from a purely sedimentological viewpoint. Allen (1967b, p. 265) has lately questioned the accepted source area of certain deltaic sediments.

Wooldridge and Linton (1955) wrote on the geomorphology of the region, with some reference to the Tunbridge Wells district; papers by Bird (1956; 1958) on the upper Medway catchment area relate more specifically to it.

The Geological Survey Wartime Pamphlets and earlier Water Supply Memoirs have been supplemented by a Well Catalogue Series of Water Supply Papers (Harvey and others 1964).

Straker (1931) made a thorough investigation into the former Wealden iron industry and his book gives details of many of the sites of old minepits, furnaces and forges. C.R.B., R.A.B., R.W.G., E.R.S.-T.

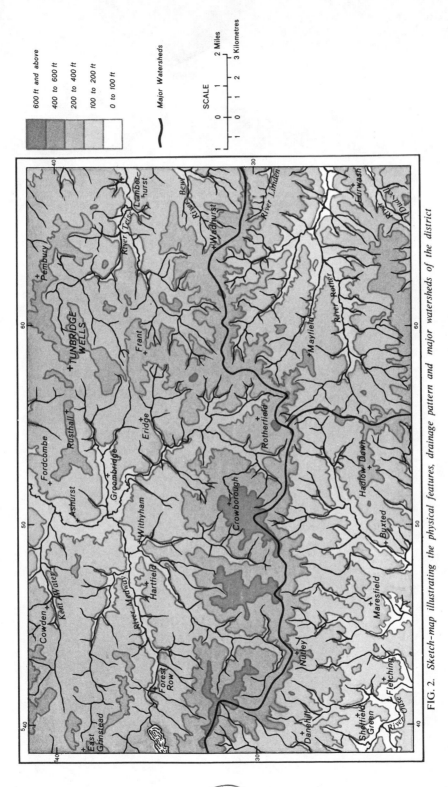

FIG. 2. *Sketch-map illustrating the physical features, drainage pattern and major watersheds of the district*

Legend:

600 ft and above

400 to 600 ft

200 to 400 ft

100 to 200 ft

0 to 100 ft

Major Watersheds

SCALE

Chapter II

STRUCTURE

GENERAL ACCOUNT

THE DISTRICT is situated in the centre of the Wealden anticlinorium, which is superimposed on a Mesozoic basin. The dominant structural elements are the Crowborough Anticline and major east to west trending strike faults (see Fig. 1 and Plate II). Folds are well developed only in the south-east, adjustment to earth movements having been mainly by faulting. The Crowborough Anticline trends west to east and dies out, being replaced by faulting in the east. Part of the northern limb of another large east to west trending anticline, the Brightling Anticline, in which the oldest surface strata of the Weald, the Purbeck Beds, are exposed, is present on the southern margin of the area. This anticline fades out to the west, and these two major folds are thus *en échelon*. A similar relationship is exhibited on a smaller scale by the South Mayfield Anticline, South Stonegate Syncline, North Mayfield Anticline and Wadhurst Park Syncline lying between the Crowborough and Brightling anticlines. Periclinal folds arranged *en échelon* have been cited by Wooldridge and Linton (1955) and Howitt (1964) as typifying the structure of the central Weald.

Strike faults are common, and have throws of up to at least 750 ft (228·6 m). The main faults are summarized in Table 1 (see also Fig. 1 and Plate II) and supplementary details are given below (p. 8). There is evidence that some of these faults are reversed, generally hading to the south (pp. 14, 15). Commonly these strike faults, usually 1 to 2 miles (1·6–3·2 km) apart, occur in complex belts and divide the area into a series of small horsts and grabens (see Fig. 1 and Plate II).

The difference between the style of deformation of the Hastings Beds and of the Weald Clay is possibly due to the Hastings Beds being more competent than the Weald Clay, which latter has instead small folds impressed upon it (see Thurrell, Worssam and Edmonds 1968, pl. II). It is also possible that proximity to the main Wealden Anticline has induced large-scale accommodation fractures. From Fig. 1 it can be seen how the crest of the Crowborough Anticline is but little affected by major faulting at its eastern or western ends, and how the fault belts trend east to west on either side of the anticline. Faults are also absent at the western end of the anticline on the adjacent Horsham (302) Sheet. It must be noted that no surface evidence was found of the fault postulated very tentatively by Falcon and Kent (1960, p. 9) in the Crowborough area. There is evidence for reversed faulting affecting the Lias of the Brightling Borehole (Falcon and Kent 1960, p. 12).

Generally the structures are those that would be formed owing to movement from the south. From evidence outside this area it is likely that these movements were part of the Alpine orogeny which reached its peak during the Oligocene–

TABLE 1

Summary of the main faults on the Tunbridge Wells (303) Sheet

Name of fault	Grid reference of locality after which fault is named	Maximum downthrow in feet, and direction	Grid reference of maximum downthrow	Notes
Barelands	609 352	100 (30·5 m) N.	635 344	
Beeches	435 413	120 (36·6 m) N.W.	443 433	
			(on Sevenoaks (287) Sheet)	
Blockfield	420 400	350 + (106·7 m) N.[1]	405 403	
Burnt Oak	514 270	300 (91·4 m) N.	490 264	
Burwash Common	640 236	550 (167·6 m) N.	655 236	Reverse fault
Button's Farm	563 234	300 (91·4 m) N.	535 231	
Buxted	489 227	60 (18·3 m) E.	495 237	
Chingley	690 334	100 (30·5 m) E.	681 338	
(on Tenterden (304) Sheet)				
Coneyburrow	625 374	300 + (91·4 m) E.	630 377	
Crowpits	529 225	200 (61·0 m) N.	520 223	Reverse fault
Danehill	402 276	450 (137·2 m) S.	388 279	
		(on Horsham (302) Sheet)		
Fletching	428 235	100 + (30·5 m) N.	440 240	
Flimwell	715 312	250 (76·2 m) S.	676 316	
Gotwick Farm	418 395	220 + (67·1 m) S.	395 399	
Greenhurst	507 251	200 (61·0 m) S.	506 251	
Hendall Wood	480 250	150 + (45·7 m) N.	498 250	Possibly the eastwards continuance of the Fletching Fault
Homestall	421 374	300 + (91·4 m) S.	432 378	
Hook Green	655 359	25 (7·6 m) S.	673 353	

TABLE 1 (continued)

Name of fault	Grid reference of locality after which fault is named	Maximum downthrow in feet, and direction	Grid reference of maximum downthrow	Notes
Kilndown	700 353	250 (76·2 m) S.	727 347	(on Tenterden (304) Sheet)
Limden	672 292	250 (76·2 m) N.	675 291	
Maresfield	466 240	50 (15·2 m) S.W.	470 238	
North Charlwood	393 342	220 + (67·1 m) S.	405 342	
North Groombridge	530 375	650 + (198·1 m) N.	580 377	
North Mayfield	586 270	200 (61·0 m) S.	607 271	
Rockrobin	624 329	50 (15·2 m) N.	645 331	
Rotherfield	557 297	70 (21·3 m) S.	560 293	
Sheffield Forest	420 262	350 + (106·7 m) S.	410 264	
Silverlands	538 333	300 + (91·4 m) S.	541 332	Reverse fault
Sleeches	507 260	150 (45·7 m) S.	510 260	
Snape	626 302	150 (45·7 m) N.	631 308	
South Charlwood	393 342	360 + (109·7 m) N.	397 340	
South Groombridge	530 375	480 + (146·3 m) N.	617 368	Continues as the Finchcocks Fault on Tenterden (304) Sheet
South Mayfield	586 270	150 (45·7 m) N.	590 269	
Stonehurst	431 406	180 + (54·9 m) W.	435 408	
Tunbridge Wells	584 392	180 (54·9 m) S.	570 390	
Withyham	498 358	350 + (106·7 m) N.	506 349	Continues as the Ticehurst Fault on Tenterden (304) Sheet

¹As many of the larger faults involve the Ashdown Beds, in which it is difficult to ascertain precise stratigraphical horizons, the figures for the downthrow are in many cases minimum ones only.

B

Miocene periods, but it has been suggested that a pre-Upper Cretaceous age is not improbable for some of the subsidiary structures in the Weald (Terris and Bullerwell 1965, p. 242).   C.R.B., R.A.B., R.W.G., E.R.S.-T.

DETAILS

FOLDS

Crowborough Anticline. The Crowborough Anticline is a fault-bounded pericline measuring approximately 12 by 6 miles (19 by 9·6 km), bringing to the surface the Ashdown Beds in the area from which they derive their name, Ashdown Forest. Two structural highs are recognized on the anticlinal crest: a lower one around Wych Cross [320 420] and a higher one near Crowborough. The highest point of the High Weald, Crowborough Beacon, 792 ft (241·4 m) O.D. [5125 3070] is situated close to the highest point of the axis, although Falcon and Kent (1960, p. 9) place the seismic structural high of the Great Oolite about 1 mile (1·6 km) S. of this locality. It is possible that the Crowborough Anticline is not just a single fold, but two minor ones *en échelon.*

Deep borings in the search for oil (see p. 16) were located close to the eastern high of the anticline and although oil was found in the Lias it was not in sufficient quantity to be of economic importance. Falcon and Kent (1960, p. 9) showed that the crest of the anticline is displaced south-eastwards at depth. Reeves (1948, p. 260, figs. 32, 33) located the structural high near Wych Cross by contouring the base of the Ashdown Beds.   C.R.B., R.A.B., R.W.G.

Brightling Anticline. The axis of this anticline trends east to west just off the southern margin of the map area, with only part of the northern limb exposed on the Tunbridge Wells Sheet. Near the core of this anticline, at the highest structural point of the Weald, the Purbeck Beds crop out. The strata at the surface are mainly Ashdown Beds, which are generally dipping at about 4°N. (See Plate II). The structure is complicated by faulting (see Anderson and Bazley 1971), but the anticline appears to die out westwards, thus being complementary to the Crowborough Anticline with which it is *en échelon.*   R.A.B.

Finch Green Syncline. This is a minor syncline whose axis lies off the northern edge of the sheet on the adjacent Sevenoaks (287) Sheet.

Only the gently dipping (3°) southern limb is seen on the present sheet. In the centre of the syncline are preserved two small inliers of Upper Tunbridge Wells Sand, while the flanks are composed of Upper Grinstead Clay, Cuckfield Stone and Lower Grinstead Clay. Because of the low dips the Cuckfield Stone has a large outcrop on the dip slopes, particularly on the Sevenoaks (287) Sheet (see Dines and others 1969).   C.R.B.

Wadhurst Park Syncline, North Mayfield Anticline, South Stonegate Syncline and South Mayfield Anticline. These are all minor and complementary structures trending approximately east to west (Plate II). The strata generally dip at less than 5° from the fold axes and the structures may be accentuated by the effects of cambering. The South Stonegate Syncline can be traced on to the Tenterden (304) Sheet.   R.A.B.

FAULTS

Beeches Fault [432 413]. This fault is responsible for the 25° dip at 300° [4382 4239] some 300 yd (274 m) S. of Greybury, and the small anticlinal fold [4417 4307] 100 yd (91 m) S.S.W. of Clatfield, on the adjacent Sevenoaks (287) Sheet (see Dines and others 1969).

Stonehurst Fault [431 406]. The vertical fault plane is exposed in the stream bank [4365 4105], 350 yd (320 m) S.E. of the Beeches, where Wadhurst Clay is thrown down to the south against Ashdown Beds. The sinuous nature of this fault is indicative of a high angle of hade.

Blockfield Fault [420 400]. This major fault throws Tunbridge Wells Sand down against Ashdown Beds over most of its length, and because of the lithological similarity of these beds its exact position can only be ascertained where the Grinstead Clay abuts against the Ashdown Beds. The minor anticline [4412 3972] in Mill Wood, 300 yd (274 m) N.E. of St. Stephen's Church, Hammerwood, is probably related to this fault; flaggy sandstones of the Ashdown Beds dip 30°N. and 60°S.

Kent Water Complex [440 401]. This fault belt is bounded by the Kent Water Fault in the north and the Blockfield Fault to the south. On the north side of the Kent Water Fault the strata dip south-south-westwards towards the fault plane. A dip of 15° in this direction was measured in the overflow channel [436 404] from the now dry Goudhurst Pond. A number of dip faults are associated with this structure [436 405 and 448 400]. The structural attitude of the beds between this fault and the Gotwick Farm Fault cannot readily be determined, but appears to be roughly horizontal. Small inliers of Wadhurst Clay in the valley bottom [437 401 and 4460 3994] may be attributable to valley bulging. C.R.B.

Herontye Inlier [400 372]. Two small complementary faults running from near Herontye [4011 3749] to Boyles Farm [3948 3629] bound a horst of almost horizontal Ashdown Beds capped in part by Wadhurst Clay. Steeper dips occur adjacent to the boundary faults and a small disused quarry [4011 2742] 250 yd (229 m) N.E. of Herontye shows siltstones and sandstones dipping at 12°S.E. Close by and parallel with this horst a small graben [404 376], $\frac{1}{4}$ mile (0·4 km) N.W. of Brockhurst, faults Lower Tunbridge Wells Sand down against Wadhurst Clay. A prominent spring line marks the outcrop of the faulted sand–clay interface along the western boundary.

Mill Place Fault [395 351]. Although an important fault on the Horsham (302) Sheet, the line of the Mill Place Fault [396 351] cannot here be placed with certainty where it brings high Wadhurst Clay against low Wadhurst Clay. Its intersection with the Ashdown Beds–Wadhurst Clay boundary now lies beneath Weir Wood Reservoir [400 350].

Gotwick Farm Fault [418 395]. A stream section [3958 3960] 500 yd (457 m) S.W. of Swite's Wood shows thickly bedded sandstones of the Lower Tunbridge Wells Sand cut by a small fault. At this point the displacement across the fault is small, but to the north [392 399], where it meets the Gotwick Farm Fault, it brings Wadhurst Clay against Lower Tunbridge Wells Sand. R.W.G.

The old pit [4060 3975] 200 yd (183 m) N.E. of the Larches exposed 6 ft (1·8 m) of disturbed flaggy, dark brown, silty, fine-grained sandstone. The sandstone beds in the centre of the pit are vertical, while in the northern part of the pit the beds dip 20°N. and in the southern part, 15°S. Wadhurst Clay crops out immediately to the south of the pit and this small anticline is presumably related to the fault. C.R.B.

Charlwood Complex [393 342]. Two major faults, the North and South Charlwood faults, bound a complex graben containing Wadhurst Clay and Tunbridge Wells Sand. The South Charlwood Fault has a sinuous outcrop and appears to dip northwards at about 50° to 60°; its maximum proven throw of about 360 ft (109·7 m) occurs 550 yd (503 m) S.E. of Charlwood [3972 3393], where it faults Ardingly Sandstone against Ashdown Beds.

The North Charlwood Fault is a steeply dipping normal fault which, at its observable maximum, brings Lower Tunbridge Wells Sand against Ashdown Beds. The throw here is probably comparable with that of the South Charlwood Fault, but it is difficult to be sure of this since the precise horizon affected within the Ashdown Beds cannot be ascertained. The plane of this fault is exposed in the stream bed [4090 3416] 450 yd (411 m) E. of Mudbrooks Farm, where Wadhurst Clay is faulted against Ashdown Beds. Eastwards of this locality the positions of both boundary faults are clearly picked out by the sharp change in vegetation where the light sandy soil of the Ashdown Beds outcrop contrasts with the heavy clay soil of the Wadhurst Clay. Within the graben several subsidiary faults, sub-parallel to the boundary fault, divide the strata into long, narrow blocks running roughly east to west. Where seen, dips within these blocks are generally into the fault planes. At Charlwood, sections in the drive [3933 3421] show siltstones and sandstones of the Lower Tunbridge Wells Sand dipping approximately northwards at 15°, 20° and 25°. At 20 yd (18 m) S. of this point these beds have a thrusted contact with the Wadhurst Clay; the outcrop of this latter is marked by a series of deep pits where the clay has been formerly worked for marl. Some 80 yd (73·2 m) N. a normal fault brings the sandstones once more into contact with Wadhurst Clay. Within the same fault block a small disused quarry [4040 3419] 120 yd (110 m) W. of Mudbrooks Farm exposes Ardingly Sandstone dipping southwards at 7°.

A deep marl pit [4029 3396], formerly worked for Wadhurst Clay, 350 yd (320 m) S.W. of Mudbrooks Farm, has been dug up to the South Charlwood Fault. Backfilling and landslip have obscured any section which might formerly have exposed the fault plane itself.   R.W.G., C.R.B.

Tunbridge Wells Fault [584 392]. The Hawkenbury Graben is bounded on the north by the Tunbridge Wells Fault and to the south by the North Groombridge Fault. The best evidence for displacement along the fault is to the east of Tunbridge Wells, where the Tunbridge Wells Sand is faulted down to the south against Wadhurst Clay. Westwards the line of the fault cannot readily be traced beneath the built-up part of the town, but small inliers of Wadhurst Clay, shown on the map ¼ mile (0·4 km) S.S.E. and 700 yd (640 m) S.S.E. of Rusthall House [5585 3875 and 5635 3900] and at Tunbridge Wells Central Station [584 392], appear to be cut off on the north side of the fault.

North and South Groombridge faults [560 380 and 560 359]. Two of the largest faults proved in the central Weald are those bounding the Groombridge Horst. Over much of the length of the faults Upper Tunbridge Wells Sand is faulted down to the north and south against Ashdown Beds. Since the exact stratigraphical position of the Ashdown Beds cannot be ascertained at the fault contact, only a minimum figure for the throw of the fault can be calculated.

A borehole at Tunbridge Wells Station is recorded in detail by Whitaker and Reid (1899, pp. 34–5) and although the exact position of the base of the Lower Tunbridge Wells Sand is uncertain the base of the Wadhurst Clay is clear and lies at −273·5 ft (83·4 m) O.D. An examination of borehole specimens suggests that the reversed fault mentioned by Harvey and others (1964, 303/3[1]) does not exist and that a normal succession to the Ashdown Beds was proved. Some 1100 yd (1006 m) S. of this borehole, just south of the North Groombridge Fault, the highest point [582 374] of the Ashdown Beds on the Groombridge Horst is approximately 480 ft (146·3 m). From Plate II it can be seen that the −125 ft (38·1 m) contour for the base of the Ashdown Beds would intersect the fault at this point. The North Groombridge Fault would here have a downthrow in excess of 600 ft (182·9 m). At least 500 ft (152·4 m) of Wadhurst Clay, Lower Tunbridge Wells Sand, Grinstead Clay and Upper Tunbridge Wells Sand and an unknown amount of Ashdown Beds are absent on the south side of the fault. A minimum figure of 440 ft (134·1 m) is indicated for the amount of downthrow of the South Groombridge Fault to the south-east of Tunbridge Wells.

---

[1]Numbers refer to the well records in the Hydrogeological Department of the Institute of Geological Sciences.

A spring line is found along the line of the fault [5510 3805], 600 yd (549 m) S. of Holmewood House. At the eastern end of the horst a cross fault (Coneyburrow Wood Fault [630 377]) lets down Tunbridge Wells Sand, and in places Wadhurst Clay, between the two main faults. This has the effect of reversing the direction of throw along the northern fault between Dundle [632 382] and Lamberhurst [680 362].

C.R.B.

The North and South Groombridge faults unite in the Teise Valley, east of Lamberhurst, continuing thence as the Finchcocks Fault, which throws down the folded Wadhurst Clay and Tunbridge Wells Sand on the south, against beds 100 ft (30·5 m) below the top of the Ashdown Beds; the displacement is approximately 200 ft (61·0 m).

E.R.S.-T.

The two tributary streams flowing to the north and south of Groombridge, which join the River Medway west of the village, appear to be fault-guided for part of their length. The line of the fault can be fixed at Warren, where the spectacular Eridge Rocks (Ardingly Sandstone) terminate abruptly [5537 3614]. To the north of Frant an inlier of Wadhurst Clay has been isolated by the fault in Chase Wood [590 362].

C.R.B.

Kilndown Fault [670 357]. This fault runs roughly parallel to the South Groombridge Fault from Spray Hill [679 355] westward towards Bayham Abbey ruins [650 363]. Its throw is difficult to determine in this area, as Tunbridge Wells Sand occurs on both sides. Near Furnace Farm [666 356], displacement of a clay seam within the formation suggests a downthrow of approximately 50 ft (15·2 m) to the N.N.E. Recent water bores at Furnace Farm (p. 70) reveal a structure more complex than that suggested by the surface outcrops. The apparent duplication or thickening of the Wadhurst Clay in the No. 2 Borehole may indicate the occurrence of a fault slice of this formation associated with the Kilndown Fault or a steepening of dip due to terminal bending of strata towards the fault. Reversed faulting is difficult to reconcile with the surface evidence. It is apparent, however, that the Kilndown Fault is a more important structure than its modest throw of 50 ft (15·2 m) hereabouts would suggest.

Hook Green Fault [655 359]. Likewise traced westward from the Tenterden area, the Hook Green Fault follows a sinuous sub-parallel course 200 to 600 yd (183–549 m) S. of the Kilndown Fault from Spray Hill through Lamberhurst Down and Hook Green, uniting with the latter 280 yd (256 m) S.S.E. of Bayham Abbey ruins. It displaces a clay seam in the Tunbridge Wells Sand, and can be shown to have a small southerly downthrow of about 25 ft (7·6 m) at Lamberhurst Down which decreases westward to practically nil at its junction with the Kilndown Fault.

Chingley Fault [660 348]. The northern edge of the Wadhurst Clay inlier which floors the site of a proposed reservoir in the Bewl Valley near Old Forge Farm [679 339] is truncated by the Chingley Fault. The structural picture is complicated by an anticline south of the fault and also by cambering and valley bulging (see Shephard-Thorn and others 1966, pp. 20–1, fig. 5). Disregarding the effects of terminal bending of strata towards the fault plane, a downthrow to the north of 75 to 100 ft (22·9–30·5 m) is indicated. Westward of the Bewl Valley the fault was traced for a further 2 miles (3·2 km), through Markwicks and Owls Castle, by its truncation of the outcrops of a clay seam within the Tunbridge Wells Sand.

E.R.S.-T.

Barelands–Rockrobin fault zone [630 340]. Between these two faults is a complex zone of relatively minor faulting (see Fig. 1), trending generally east to west and cut off in the west by a minor fault trending north-east. None throws down more than 100 ft (30·5 m) and a series of minor horsts and grabens is formed, with the Tunbridge Wells Sand and Wadhurst Clay repeated. The faults are mainly traced by mapping thin clay seams that occur within the Tunbridge Wells Sand. The sinuous nature of the minor fault near Lightlands [5965 3335] indicates a fault with a relatively high angle of hade.

R.A.B.

To the south of the anticlinal axis referred to above (see Chingley Fault) the strata of the Bewl Valley Inlier have a general south to south-easterly dip, but numerous faults complicate the structure. The Barelands–Rockrobin fault zone extends into the Cousleywood [652 334] area, where a thin silty clay seam about 50 ft (15·2 m) above the base of the Tunbridge Wells Sand is repeated three times within 100 yd (91 m) by two faults throwing down southerly and trending slightly north of east. A set of three parallel faults trending east-south-east extends from Cousleywood towards Bryant's Farm [666 328], displacing the Tunbridge Wells Sand–Wadhurst Clay junction. They are met by a further set trending south-east including the Kilndown and Tice-hurst faults extending from the Tenterden district. Between these strike faults, short conjugate faults at about 060° demarcate small inliers of Ashdown Beds ¼ mile (0·4 km) S. of Lower Hazelhurst [677 321] and near Lower Tolhurst [673 310]. At Pinton Hill, some 200 yd (183 m) S. of Rowley [679 315], a slice of Wadhurst Clay is brought up by the unnamed fault south of the Flimwell Fault.   E.R.S.-T.

Withyham Fault [498 358]. One of the longest faults in this area can be traced along a sinuous course from Thornhill [4310 3725] in the west 22 miles (35 km) eastwards on to the adjacent Tenterden (304) Sheet, having joined with the Ticehurst Fault south of Birchett's Green [668 313]. The fault is complementary to the South Groombridge Fault and forms the southern boundary fault to the Withyham–Frant Graben. As with the Groombridge faults the maximum throw is in places in excess of 400 ft (121·9 m).   C.R.B.

The fault divides about ½ mile (0·8 km) E. of Eridge Station [5420 3450] (Fig. 1), its displacement being taken up more or less equally by the two resulting faults. For much of its path eastwards it is associated with minor fractures trending in a similar direction. The effect of the fault is to throw down Tunbridge Wells Sand on the north against Wadhurst Clay and Ashdown Beds to the south. Springs and much seepage of water in the valley east of Danegate [5630 3383] indicate the presence of several faults. The sinuous nature of the main fault indicates that the fault plane is not vertical.   R.A.B.

Silverlands Fault [540 332]. Wadhurst Clay, Lower Tunbridge Wells Sand, Ardingly Sandstone and Grinstead Clay, inclined 4° N.N.E., terminate against this fault. They are let down against Ashdown Beds, a downthrow of at least 300 ft (91·4 m). The dip of 20° N.N.E. in the Lower Tunbridge Wells Sand in an old pit [5187 3316], 300 yd (274 m) N.E. of Gilridge Farm, is probably related to the fault. Small splinter faults occur at both the western and eastern ends of this fracture. It is possible that there is a belt of undetected faulting in the Ashdown Beds which connects the Silverlands Fault with the Charlwood faults (see Fig. 1 and Plate II).   C.R.B.

Snape Fault [630 307]. In the stream through Birchett's Wood, 200 yd (183 m) E.S.E. from Foxes Bank, rocks within the fault zone are exposed [6365 3071]. A spring line is present along the fault. In Birchett's Wood seepage from the sands to the south has caused marshy ground, and on the north there are many landslips in the clays of the valley sides. The throw along most of the fault is at least 150 ft (45·7 m), but it is of limited extent (Fig. 1 and Plate II) and dies out quickly west of Bestbeech Hill [6185 3141]. In the east the movement was probably taken up by the Limden Fault which lies on a similar line, but a little to the south.   R.A.B.

Limden Fault [670 292]. Throughout its length this fault throws Wadhurst Clay down against Ashdown Beds. Between Wedd's Farm [6810 2922] and Limden [6712 2919] the line of the fault can be accurately placed where it is marked on the northern, Wadhurst Clay side, by a line of old clay pits and on the southern side by a number of exposures of flaggy and massive sandstone (Ashdown Beds). At its maximum, near Wedd's Farm, it brings low Wadhurst Clay against Ashdown Beds some 150 to 200 ft (45·7–61·0 m) below the junction with the Wadhurst Clay.   R.W.G.

Danehill Fault [402 276] **and Sheffield Forest Fault** [420 262]. On its southern side the Crowborough Anticline is bounded in part by a series of faults which bring Ashdown Beds against Wadhurst Clay and Tunbridge Wells Sand. Eastwards from Funnell's Farm [442 266] the Sheffield Forest Fault splits into a number of shorter faults, each throwing down to the north by 200 to 300 ft (61·0–91·4 m), to form a series of fault blocks in which Ashdown Beds lie to the south of Wadhurst Clay which in turn lies to the south of Tunbridge Wells Sand. Between the Sheffield Forest and Fletching faults the strata dip regularly southwards at about 2°.   C.R.B., R.W.G.

Scaynes Hill Fault [400 233]. Near Wapsbourne Farm [398 233] this fault displaces the outcrop of the Grinstead Clay. Eastwards of this point the line of the fault is no longer traceable because of the absence of marker beds in the Upper Tunbridge Wells Sand.   R.W.G.

Fletching Fault [429 234]. The two right-angled bends of the River Ouse south-west of Fletching may be the result of stream adjustment to flow along the more easily eroded shatter zone of this fault. North-eastwards from here the fault is responsible for the shift in outcrop of the Grinstead Clay near Down Street [447 245] and the restriction of the small inlier of Grinstead Clay ¼ mile (0·4 km) W. of Down Street (Downstreet Pit 440 240) to the upthrow, south side of the fault. This fault curves around to the north-west of Maresfield, where it apparently changes its direction of downthrow, probably as a result of the Maresfield Fault, which throws down to the south-west.   C.R.B.

In the region of Lane End Common [403 221] the line of the Fletching Fault is marked by a number of strong springs. In this area there is no obvious displacement across the fault, but the beds lying to the north dip northwards towards the Scaynes Hill Fault whilst those on the south side dip southwards at a regular 1° to 2°.
     R.W.G.

Hendall Wood Fault [480 250]. This is a sinuous east to west trending, low-angled fault which dies out eastwards against the Greenhurst Fault. The downthrow along it is not constant, being complicated by a number of cross faults. To the north of the fault occurs a triangular fault-bounded horst (Hendall Horst) of Ashdown Beds: in the south the Hendall Wood Fault throws Wadhurst Clay down against the Ashdown Beds; to the north-west Ardingly Sandstone abuts the Hendall Horst; while to the north-east a splinter from the Greenhurst Fault throws Wadhurst Clay and Lower Tunbridge Wells Sand down against the Ashdown Beds. Between the Green hurst and the Hendall Wood faults a small north to south cross fault [473 250] lets down the Upper Tunbridge Wells Sand to the east. The net effect is that here the Hendall Wood Fault has a downthrow of approximately 80 ft (24·4 m) N. In the report of the borehole [493 248] at Warminghurst (Whitaker and Reid 1899, p. 17; Harvey and others 1964, 303/16), which was sunk 65 ft (19·8 m) S. of the fault, the total thickness given to the Wadhurst Clay is only 147 ft (44·8 m) and it is probable that the base of the Wadhurst Clay has been cut out by the southward dipping Hendall Wood Fault. The fault plane is calculated to dip 71°S. At Dallings [4834 2498], just over ½ mile (0·8 km) W.N.W., 172 ft (52·4 m) of Wadhurst Clay were recorded (op. cit., 303/17), and in two wells at Uckfield on the adjacent Lewes (319) Sheet thicknesses of 182 and 187 ft (55·5, 57·0 m) were recorded by Whitaker and Reid (1899). A dip of 5°N. in the road cutting [4932 2483] immediately north of Warminghurst may be the result of terminal flexuring associated with this fault. The triangular block of country to the south of the Hendall Wood Fault, with the Maresfield Fault to the south-west and the Buxted Fault to the south-east, has strata with a regional dip of about 1°S.   C.R.B.

Burnt Oak Fault [510 271]. In the stream [5205 2729], 750 yd (686 m) at 068° of Burnt Oak, there are grey clays and siltstones of the Wadhurst Clay and just a few yards downstream, across the fault plane, massive sandstones and silts of the Ashdown Beds are found. Seepage, causing lines of reeds, gives evidence of the position of the fault on the nearby hillsides, as it does also farther west near High Hurstwood [4832 2645] where Grinstead Clay abuts against Ashdown Beds.   R.A.B., C.R.B.

Sleeches Fault [506 260]. The sinuous nature of this east to west trending fault especially in the region of Sleeches Farm [5070 2600] and 400 yd (366 m) S. of Ford Brook [5178 2675] is indicative of a fault plane hading to the south.   R.A.B.

Greenhurst Fault [500 252]. A branch of this fault can easily be traced in the bed of the stream which flows through Stonehouse Wood [482 255], for the stream repeatedly meanders across the fault plane to expose thin flaggy sands and silts of the Ashdown Beds on the south-west, upthrow, side of the fault and thin silts and shales of the Wadhurst Clay on the north-east, downthrow, side.   C.R.B.

West of Howbourne Farm [5160 2500] the fault causes a marked east to west scarp with the harder Tunbridge Wells Sand beds on the south standing well above the Wadhurst Clay beds to the north. The throw is probably at least 200 ft (61·0 m). In the east the fault dies out, but it is possible that a minor extension trends to the north-east to join the Mayfield fault zone.

Rotherfield Fault [560 293]. This relatively minor strike fault, with a maximum throw of about 70 ft (21·3 m) to the south, trends E.N.E. to W.S.W. in an area otherwise free from complications due to faulting. In the west it fades out within 1 mile (1·6 km). Similarly it fades out to the east, no indication of its course being found in the Ashdown Beds. The relative lack of faulting around the nose of the Crowborough Anticline is noted on p. 5.

Mayfield faults [585 270]. The North Mayfield Fault has a throw of at least 200 ft (61·0 m) along most of its length. Where sands and silts of the Tunbridge Wells Sand are thrown down against sands and silts of the Ashdown Beds this fault, like many others in the area, is difficult to trace with precision. West of Mayfield, near Freeman's Farm [5625 2690] a cross fault, ½ mile (0·8 km) in length and throwing down to the east, connects the North Mayfield Fault to the Sleeches Fault.

In Mayfield itself the fault was proved by a temporary ditch section, only 3 ft (0·9 m) deep, along the main A267 road just north of the Roman Catholic Convent [5879 2725]. The section started in sands and silts of the Tunbridge Wells Sand on the south of the fault, and northwards went into stiff grey clays with thin '*Cyrena*' limestones and ironstone beds. Faunal evidence indicated that the latter were from the lower part of the Wadhurst Clay. In a roadside section opposite the convent and just south of the fault minor faults are exposed in Ashdown Beds sandstones and silts. These are associated with the main fault.

To the east the fault is traced by a line of springs giving rise to marshy ground with reeds in places, particularly between Rolf's Farm [6145 2710] and Forge Farm [6400 2670]. In this area to the north of the fault the strata dip generally southwards, and to the south generally northwards. The fault thus marks the axis of a synclinal structure. A minor syncline, probably formed at the same time as the fault which took up most of the pressure, is found just to the south (see Fig. 1 and Plate II) of this fault. This syncline, proved near Golds Farm [6300 2686], apparently extends eastwards (South Stonegate Syncline) but is only a minor structure the effect of which in this area is accentuated by faulting. The North Mayfield Fault can only be traced for 700 yd (640 m) E. of Forge Farm [6400 2670] and is then lost beneath the alluvium of the River Rother, and cannot be traced across a large area of Wadhurst Clay to the east. It is probable that, although the throw is diminishing eastwards, the fault continues in a generally eastward direction and is represented by a small fault, with a throw of about 50 ft (15·2 m), 500 yd (457 m) at 145° from Battenhurst Farm [6750 2705] (see Plate II).

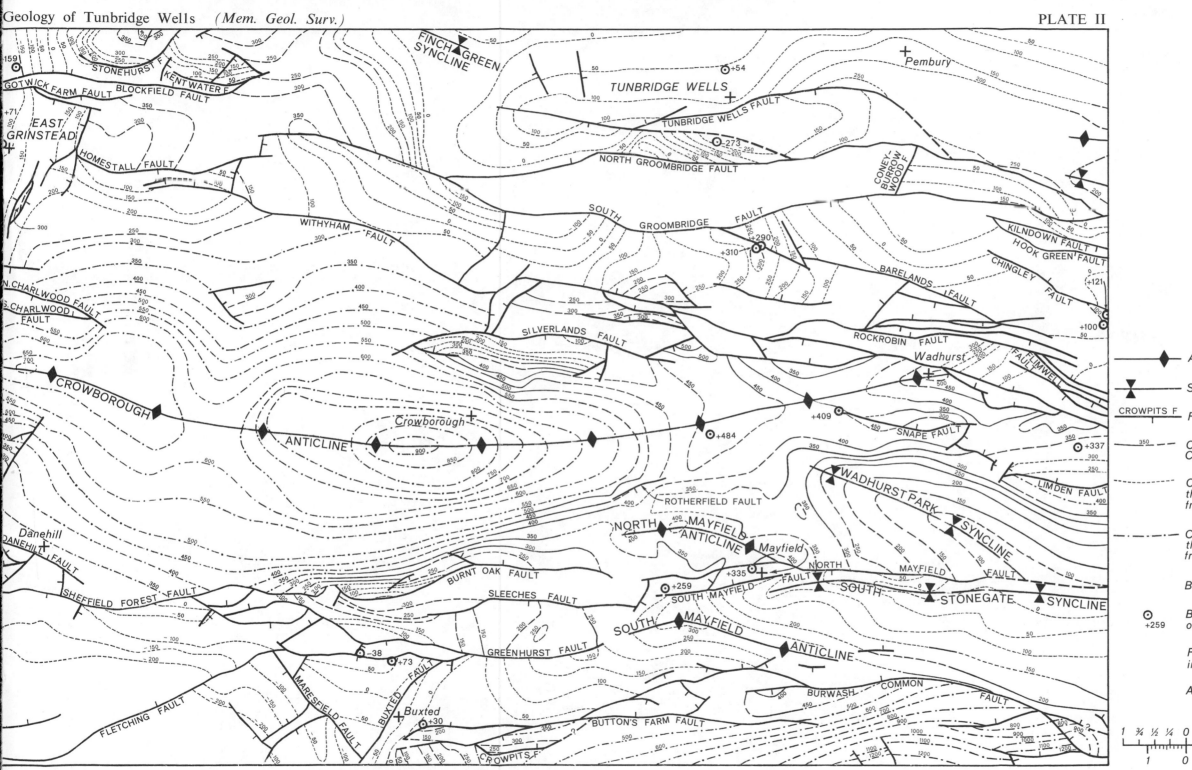

SKETCH–MAP OF THE STRUCTURAL GEOLOGY OF THE TUNBRIDGE WELLS AREA.

INDEX

◆——— Anticlinal axis

▼——— Synclinal axis

CROWPITS F ——— Fault, crossmark on the downthrow side

——— 350 ——— Contour drawn on the base of the Wadhurst
Clay, figure indicates feet above Ordnance Datum

- - - - - - - Contour drawn on the base of
the Wadhurst Clay extrapolated
from overlying formations

– – – – – Contour drawn on the base of
the Wadhurst Clay extrapolated
from underlying formations

Broken lines denote estimation

⊙
+259   Borehole showing height in feet
of the base of the Wadhurst Clay

Figures preceded by + or – indicate levels
in feet above or below Ordnance Datum

Abbreviation : F . . . FAULT

SCALE

The South Mayfield Fault forms the southern confine of the graben on which Mayfield village is situated. This is a relatively minor fault, but is the cause of a spring line and of the inlier of Wadhurst Clay just south of Mayfield.

Burwash Common Fault [640 241]. To the south of Buxted this fault ends in the Wadhurst Clay (Fig. 1). In the area of Wilderness Farm [5340 2316] it has a throw of at least 250 ft (76·2 m) to the north. Between Old Mill Farm [5876 2450] and Laurelhurst [6553 2351] this may exceed 500 ft (152·4 m) (see Plate II). The displacement decreases to the west and increases eastward. There may be other minor faults cutting the Ashdown Beds between Broadoak and Burwash Weald, but none could be traced owing to a lack of mappable horizons.

The fault branches south of Ryegreen [6640 2349], and in High Wood a borehole [6654 2314] (303/179, manuscript record in the Institute of Geological Sciences) proved that the Purbeck Beds are repeated and that the northern branch is a reverse fault. The normal thickness of Purbeck Beds in this region is about 460 ft (140·2 m), but here it is recorded as 828 ft (252·4 m). The southerly branch is possibly also a reverse fault, as about 130 yd (119 m) beyond the southern margin of this district, in a south-easterly direction, is another borehole (Falcon and Kent 1960, p. 10; Cole and others 1965, 319/309) in which reverse faulting is proved in Liassic strata at 3600 ft (1097·3 m). It has been suggested (Howitt 1964) that this reverse fault never reaches the surface in this area, but dies out in the Kimmeridge Clay. Whilst its throw is possibly diminished it seems more likely that this major fault is represented at the surface by one of the branches through High Wood of the Burwash Common Fault.

From the sinuous outcrop of this fault it is likely that it is a reverse fault with a hade to the south all along its length.

Button's Farm Fault [565 233]. Associated with the Burwash Common Fault this fault may extend farther eastwards than shown, but it could not be traced through the area of continuous Ashdown Beds outcrop. Generally it is traced by lines of seepage along the fault plane. In much of the area the fault plane has been etched out by erosion causing slight features, e.g. just south of Herring's Farm [5765 2335] where the line of the fault is indicated by a small nick in the ridge.

Crowpits Fault [530 226]. Again this is associated with the Burwash Common Fault (Fig. 1). A borehole [5201 2227] (Harvey and others 1964, 303/99), 630 yd (576 m) at 085° of the Smithy, Shepherd's Hill, is now interpreted as 65 ft (19·8 m) Ashdown Beds on 133 ft 6 in (40·7 m) Wadhurst Clay on 51 ft 6 in (15·7 m) Ashdown Beds. This indicates a reverse fault hading at about 45°S. Reeds indicating seepage along the fault are abundant down the sides of the valley between Shepherd's Hill and Crowpits.

This reverse faulting, like the Burwash Common Fault, is in accord with the theory of pressures coming dominantly from the south at the time of the earth movements that formed the structures of the area. R.A.B.

Chapter III

CONCEALED FORMATIONS

GENERAL ACCOUNT

WITHIN THE confines of the Tunbridge Wells Sheet there is only one deep borehole, Ashdown No. 2 of D'Arcy Exploration Company Limited, that penetrated the thick concealed Mesozoic strata to enter the Palaeozoic; the adjacent borehole, Ashdown No. 1, stopped in the Lias (see p. 22). There are, however, a number of boreholes just outside this district in which Palaeozoic rocks have been proved. These are: Penshurst (1938), 1½ miles (2·4 km) to the north (Falcon and Kent 1960; Dines and others 1969); Brightling, 120 yd (110 m) beyond the south-east corner of the map; and Grove Hill, Hellingly, 5 miles (8 km) to the south of the map (Falcon and Kent 1960). A fourth borehole, at Heathfield, 400 yd (366 m) south of the sheet boundary, only just entered the Kimmeridge Clay (manuscript record in Institute of Geological Sciences)[1].

From these deep boreholes it is possible to picture the shape of the Jurassic basin and also the changes in thickness and lithology which have been observed across the basin (see Shephard-Thorn and others 1966, p. 11). By reference to the isopachyte map of Gallois (1965a, p. 13) it can be seen that the Tunbridge Wells district is centrally situated in the Wealden Basin, lying to the south of the London Platform.

Allen (1954; 1960b; 1961; 1967b), in work based on analyses of pebbles and heavy mineral suites, has attempted to reconstruct the geology of the London Uplands as it existed during Hastings Beds times.

C.R.B., R.A.B., R.W.G., E.R.S.-T.

DETAILS

STRATIGRAPHICAL PALAEONTOLOGY OF THE ASHDOWN BOREHOLES

Lithological descriptions and basic data on these boreholes are given in Appendix I. In Ashdown No. 1 Borehole the following approximate depths (in feet) were more or less fully cored: 660–1225 (201·2–373·4 m), 1696–7 (516·9–517·4 m), 2083–6 (634·9–635·8 m), 2417–27 (736·7–739·7 m), 2471–2759 (753·2–840·9 m), 2926–3192 (891·8–972·9 m), 3242–60 (988·2–993·6 m), 3402–56 (1036·9–1053·4 m), 3664–76 (1116·8–1120·4 m), 3722–3830 (1134·5–1167·4 m), 3965–4014 (1208·4–1223·5 m), 4034–54 (1229·6–1235·7 m), 4100–16 (1249·7–1254·5 m), 4481–8 (1365·8–1367·9 m) and 4509–35 (1374·3–1382·2 m). In Ashdown No. 2 Borehole the approximate depths (in feet) which were more or less fully cored were: 1112–90 (338·9–362·7 m), 1220–40 (371·9–378 m), 2910–45 (887–897·6 m), 3620–90 (1103·4–1124·7 m), 4757–62 (1449·9–1451·5 m), 5270–300 (1606·3–1615·4 m), 5533–70 (1686·5–1697·7 m) and 5585–709 (1702·3–1740·1 m).

In both boreholes further information was obtained by geologists of B.P. Petroleum Development Limited from the examination of the chip samples recovered from the

[1]See Addendum on p. 24.

intervening depths. Throughout this account, and in Appendix I, depths and thicknesses are given as logged from the height of the rotary table and not from ground level. No correction has been made for dip; all the strata appear to be essentially horizontal or with no recorded dip.

We are grateful to Dr. J. H. Callomon and Dr. W. G. Chaloner for identification and zonation of most of the Upper Jurassic ammonites and of Lower Jurassic–Triassic? spores respectively. Dr. F. W. Anderson has dealt with the ostracods of the Wealden–Purbeck Beds and Dr. R. Casey provided identifications of bivalves from the Purbeck Beds and of the Portland Beds ammonites.

As would be expected from their proximity to each other—the boreholes are only 1 mile (1·6 km) apart—there is considerable similarity in each sequence of Lower Cretaceous and Jurassic. The apparent variations found may be partly due to there being only four cored horizons in Ashdown No. 2 which are equivalent to beds cored in Ashdown No. 1 and of these only the Kellaways Beds can be correlated with precision. The other variations may be as much due to the relative lag in recovery of chip samples and to their interpretation as to actual differences in the formations. However there are consistent differences between the two records which do suggest a regional thinning of the Jurassic succession to the south, continued towards the Henfield Borehole (Falcon and Kent 1960; Chaloner 1962) and which is complementary to the thinning known farther north towards the Penshurst boreholes (Dines and others 1969) and Warlingham Borehole (Worssam and Ivimey-Cook 1971).

These differences between the Ashdown Boreholes can be tabulated:

	Ashdown No. 1		Ashdown No. 2	
	Depth Ft (m)	Interval Ft (m)	Depth Ft (m)	Interval Ft (m)
Base of Purbeck Beds	1099 (335·0)		1180 (359·7)	
		2622 (799·2)		2472 (753·5)
Top of Kellaways Beds	3721 (1134·2)		3652 (1113·1)	
		724 (220·7)		684 (208·5)
Top of Upper Lias	4445 (1354·8)		4336 (1321·6)	

Purbeck Beds. Ashdown No. 1, 479 to 1099 ft (146–355 m)—620 ft (189 m).
Ashdown No. 2, 559 to 1180 ft (170·4–359·7 m)—621 ft (189·3 m).

The stratigraphy and ostracod faunas of these beds are described in Anderson and Bazley (1971). Dealing with the mollusca of the Purbeck facies in Ashdown No. 1, Dr. R. Casey comments: "Coring commenced at 660 ft (201·16 m) in a variable-salinity sequence evidently representing the *Corbula*-Beds of the Dorset succession. Beds dominated by the bivalve *Neomiodon*, often a rock-former, with a few gastropods such as *Ptychostylus harpaeformis* (Koch and Dunker) and *Paraglauconia strombiformis* (Schlotheim), are inferred to have been laid down under conditions of very low salinity. The more brackish facies contains the bivalves *Corbula* spp., *Modiolus* sp., *Eocallista* (*Hemicorbicula*) *parva* (J. de C. Sowerby) and reduced numbers of *Neomiodon*, with the diminutive gastropod *Hydrobia chopardiana* (de Loriol) abundant at some levels. Two narrow bands at 662 and 663 ft (201·77, 202·08 m), respectively point to a near-marine environment of deposition, having yielded *Corbula*, *Protocardia* and a dwarf species of *Musculus*, together with the gastropod *Pachychilus manselli* (de Loriol). At 714 to 716 ft (217·62–218·23 m) a thicker development of quasi-marine strata is characterized by *Corbula spp.*, *Protocardia sp.*, *Grammatodon sp.* and pectiniform bivalves: possibly this horizon corresponds to the Scallop Bed of Dorset. Assemblages of brackish-water mollusca are less frequent through the subjacent 70 ft (21·33 m) of core, the beds yielding mainly *Neomiodon medius* (J. de C. Sowerby) and allied species, *Unio*, *Protelliptio* and a few *Viviparus*. This part of the succession contains 'dirt-beds' and much plant debris. A thick band of marine or near-marine strata between 785 ft and 796 ft 7 in (239·26, 242·79 m) is correlated with the Cinder Beds, which form the base of the Durlston Beds or Cretaceous part of the Purbeck Beds in Dorset. *Liostrea distorta* (J. de C. Sowerby) and *Myrene fittoni* Casey are crowded at the base and at the top of this band; abundant crushed *Protocardia major* (J. de C. Sowerby) occur in between, as in the middle part of the Dorset Cinder Beds. *Myrene* is still common at this intermediate level, but of smaller size and greater variety; *M. pellati* (de Loriol) and *M. mantelli* (Dunker) are here found together with *M. fittoni*. Other marine bivalve genera found in this band are *Isognomon*, *Modiolus* and *Isocyprina*, and the gastropod *Hydrobia chopardiana* is again well represented. Below this marine band—easily recognizable in all the relevant Wealden borings on account of its distinctive bivalve fauna—the beds are characterized by mainly fresh-water mollusca such as *Neomiodon*, *Unio* and a few *Viviparus*. A brackish band at 824 to 826 ft (251·15–251·76 m) is indicated by the presence of *Eocallista* (*Hemicorbicula*) *parva*. Few determinable molluscs were found in the Lower Purbeck. *Neomiodon* occurred at 1018 ft (310·28 m).

Preservation of macrofossils in the Durlston Beds was unusually good and the core-samples provided the holotype and paratypes of *Myrene fittoni* (Casey 1955a, pl. 11, figs. 1, 2, p. 219) and examples of *Eocallista* (*Hemicorbicula*) *parva* showing fine details of hinge structure (Casey 1955b, p. 370, fig. 6B)."

Portland Beds. Ashdown No. 1, 1099 to about 1181 ft (335–360 m)—about 82 ft (25 m).
　　　　　　　　Ashdown No. 2, 1180 to about 1250 ft (359·7–381 m)—about 70 ft
　　　　　　　　(21·3 m).

Below the anhydrite beds of the Purbeck are black silty shales and mudstones which show cross bedding, silt layers and also ripple marks. These are included in the Portland Beds with a top taken at the base of the anhydrite. The mudstones near the top are rather more calcareous and become progressively more shaly downwards. The base of these beds is poorly marked on both faunal and lithological criteria and the basal junction, like the top, is arbitrarily taken at the depth quoted.

The highest ammonites from Ashdown No. 1 are indeterminate remains at 1129 ft 6 in (344·3 m). A *Paracraspedites?* (or *Lomnosovella?*) occurred at 1147 ft 8 in (349·8 m), *Crendonites?* at 1168 ft 6 in (356·2 m) and *Epivirgatites* at 1178 ft 2 in and 1181 ft (359·1, 360 m). In Ashdown No. 2 fragments of an indeterminate giant pavloviid ammonite occurred at 1183 and 1184 ft (360·6, 360·9 m), *Epivirgatites?* with an indeterminate perisphinctid at 1226 ft 6 in (373·8 m) and another pavloviid at 1227 ft (374 m). Dr. Casey observes that, despite the occurrence of an ammonite tentatively assigned to *Paracraspedites* they could well all belong in the lower parts of the Portland Beds and there is no definite evidence of the presence of the higher parts of the formation which in the Penshurst boreholes (op. cit.) were more calcareous. The ammonites at 1226 to 1227 ft (373·7–374 m) (Ashdown No. 2) could also be assigned to the Kimmeridge Clay though this would result in a greater anomaly between the thicknesses of the Portland Beds in the two boreholes.

Kimmeridge Clay. Ashdown No. 1, about 1181 to about 3020 ft (360–920·5 m)—about 1839 ft (560·5 m).
　　　　　　　　Ashdown No. 2, about 1250 to about 2980 ft (381–908·3 m)—about 1730 ft (527·3 m).

The substantial thickness of argillaceous sediments of this formation is varied only by a number of shell bands (noted for the British Petroleum Company Limited by

M. W. Strong), and by local more bituminous mudstones and argillaceous limestones. Between 2350 and 3000 ft (716·3–914·4 m) in Ashdown No. 1, however, the sequence becomes more arenaceous with silty mudstone and frequently calcareous sandstones. Previously (Falcon and Kent 1960) these arenaceous horizons have been placed in the Corallian Beds. The lowest beds assigned to the Kimmeridge Clay revert to dark grey calcareous shale and mudstone.

The respective thicknesses of Upper and Lower Kimmeridge Clay in the boreholes have not been determined. In Ashdown No. 2 the only core specimens of this sequence were grey muddy siltstones with plant fragments at 2941 ft (896·4 m) and a shell-detrital grey limestone at 2916 ft 6 in (888·9 m). No ammonites are recorded. In Ashdown No. 1 many more specimens are available but there are major gaps in the coring.

In beds assigned to the Upper Kimmeridge Clay (Ashdown No. 1) *Pavlovia?* occurs at 1185 to 1186 ft (361·2–361·5 m) and *Pavlovia sp.* at 1195 and 1206 ft (364·2, 367·6 m) indicating over 20 ft (6·1 m) of the *Pavlovia* zones of the Upper Kimmeridge Clay. Below, indeterminate late *Pectinatites* (*Virgatosphinctoides*) were identified between 1208 and 1224 ft (368·2, 373·1 m).

In beds certainly of Lower Kimmeridgian age, *Aulacostephanus* ranged from 2085 to 2749 ft (635·5–837·9 m) with a *Pararasenia?* at 2618 ft (798 m); *Aspidoceras* from 2517 to 2712 ft (767·2–826·6 m); and *Sutneria* from 2480 to 2688 ft (755·9–819·3 m). *Amoeboceras* was found at a few levels between 2552 and 2689 ft (777·8–819·6 m).

The principal occurrences of species of *Aulacostephanus* down to 2644 ft (805·9 m) were: *A.* cf. *pseudomutabilis* (de Loriol) between 2486 and 2618 ft (757·7–798 m); *A. pseudomutabilis anglicus* (Steuer) at 2514, 2520 and 2617 ft (766·3, 768·1, 797·7 m); *A.* (*Aulacostephanoceras*) *pusillus* Ziegler at 2517 and 2519 ft (767·2, 767·8 m) and *A.* cf. *pusillus* at 2616, 2635 and 2644 ft (797·4, 803·1, 805·9 m); *A.* (*Aulacostephanoceras*) *eudoxus* (d'Orbigny) at 2628 ft (801 m) and *A.* cf. *eudoxus* at 2520 ft (768·1 m); *A.* (*Aulacostephanoceras*) *dossenus* Ziegler at 2615 and 2618 ft (797·1, 798 m) and *A.* cf. *dossenus* at 2636 ft (803·5 m). Below 2644 ft (805·9 m) much finer ribbed forms are found: *A.* (*Aulacostephanoides*) cf. *linealis* (Quenstedt) between 2645 and 2714 ft (806·2–827·2 m), *A.* cf. *circumplicatus* (Quenstedt) at 2749 ft (837·9 m); *A.* (*Aulacostephanites*) *eulepidus* (Schneid) and *A.* cf. *eulepidus* at 2699 and 2700 ft (822·7, 823 m) and commonly between 2716 and 2718 ft (827·8–828·4 m).

Sutneria was represented by two species, *S. cyclodorsata* (Moesch) at 2519 and 2688 ft (767·8, 819·3 m) and *S. eumela* (d'Orbigny) between 2480 and 2603 ft (755·9–793·4 m), with some intermediate forms around 2595 ft (791 m) and also many specimens not closely identifiable but probably also belonging to these species. The genus was particularly common between 2480 and 2496 ft (755·9–760·8 m) where it was the predominant ammonite. *Amoeboceras* (*Amoebites*) cf. *cricki* (Salfeld) occurred at 2552 and 2620 ft (777·8, 798·6 m).

Dr. Callomon observes that the species of *Aulacostephanus* suggest that the boundary between the *eudoxus* and *mutabilis* zones lies at 2644 ft (805·9 m). They and the species of *Sutneria* show the *eudoxus* Zone to extend up to at least 2480 ft (755·9 m), a thickness of at least 164 ft (50 m). *Sutneria* dominates the fauna about 150 ft (45·7 m) above the base of the zone as was also found in the Warlingham Borehole (Callomon and Cope 1971), where the thickness of the zone was 198 ft (60·4 m). Two other, indeterminate, specimens of *Aulacostephanus*, at 2085 and 2422 ft (635·5, 738·2 m) give a minimum combined thickness for the *eudoxus* and *autissiodorensis* zones of 599 ft (182·6 m), compared with 280 ft (85·3 m) at Warlingham and about 300 ft (91·4 m) in Dorset. Below 2644 ft (805·9 m) the *mutabilis* Zone extends down to at least 2749 ft (837·9 m) and possibly beyond, but there are no ammonites. Notable is the occurrence here, already in the *mutabilis* Zone, both of *Sutneria* (*S. cyclodorsata* and *S. sp.* at 2678 and 2688 ft (816·3, 819·3 m)) and *Exogyra virgula* (Defrance) at 2715 ft (827·5 m).

At Warlingham where the record was much more complete, these were found only above the base of the *eudoxus* Zone.

The level of the base of the Kimmeridge Clay can only be taken arbitrarily as there is neither good lithological nor faunal evidence. The argillaceous facies rests between 2943 and 2994 ft (897–912·6 m) on light grey calcareous sandstones capped by a thin bed of oolitic hard grey limestone, whilst below, the sequence continues as grey calcareous shale and silty shale with rare limestone bands.

In the Penshurst (1938) Borehole (Dines and others 1969) the presence of *Prorasenia* cf. *bowerbanki* Spath at 2609 to 2610 ft (795·2–795·5 m) together with some other ammonites was taken to indicate that the thicker arenaceous sequence between 2417 and 2494 ft (736·7 and 760·2 m) in that borehole belonged in the Lower Kimmeridge Clay, the base of which was taken at 2629 ft (801·3 m), i.e. some 35 ft (10·7 m) below the arenaceous beds. In Ashdown No. 1 the presence of *Ringsteadia* cf. *anglica* Salfeld at 3050 ft (929·6 m) indicates the Corallian Beds. In comparison with the Penshurst (1938) Borehole the base of the Kimmeridge Clay has been taken at 3020 ft (920·5 m) in Ashdown No. 1 at the top of a hard light grey argillaceous limestone and at a comparable horizon in Ashdown No. 2.

Corallian Beds. Ashdown No. 1, about 3020 to 3404 ft (920·5–1037·5 m)—about 384 ft (117 m).

Ashdown No. 2, about 2980 to 3348 ft (908·3–1020·5 m)—about 368 ft (112·2 m).

Apart from *R.* cf. *anglica* at 3050 ft (929·6 m) the fauna includes *Ringsteadia sp.* at 3081 and 3139 ft (939·1, 956·8 m), *Ringsteadia?* at 3138, 3142 and 3156 ft (956·5, 957·7, 961·9 m) and *Ringsteadia* or *Decipia* at 3152 ft (960·7 m), all indicating the Upper Oxfordian and equivalent to part of the Ampthill Clay. The clay facies yielded one further ammonite in Ashdown No. 1, a specimen of *Amoeboceras* (*Prionodoceras*) cf. *serratum* (J. Sowerby) at 3243 ft (988·5 m), which Dr. Callomon attributes to the *cautisnigrae* Zone.

The Corallian Beds are taken to extend to the base of a 3-ft (0·9-m) pyritic limestone recorded from 3401 to 3404 ft (1036·6–1037·5 m) in Ashdown No. 1, the lowest limestone noted in this part of the sequence. Below occur further dark grey calcareous shales which at 3433 ft (1046·4 m) contain ammonites possibly of the *cordatum* Zone, *costicardia* Subzone, i.e. near the top of the Oxford Clay.

A similar succession of argillaceous beds with a few thin limestones was recorded from Ashdown No. 2 and the boundaries are drawn to correspond. No core was taken in this part of the sequence.

Oxford Clay. Ashdown No. 1, about 3404 to 3721 ft (1037·5–1134·2 m)—about 317 ft (96·6 m).

Ashdown No. 2, about 3348 to 3652 ft (1020·5–1113·1 m)—about 304 ft (92·7 m).

The highest beds of dark grey calcareous shale overlie dark brown bituminous shelly pyritic mudstone. In Ashdown No. 1 at 3412 ft 9 in (1040·2 m) occurred a specimen of *Ochetoceras* (*Campylites*) *henrici* (d'Orbigny) with *Cardioceras* cf. *ashtonense* Arkell (or *C.* cf. *persecans* S. Buckman sp.) at 3433 ft (1046·4 m), *C.* (*Vertebriceras?*) *quadrarium* S. Buckman at 3433 ft 1 in (1046·4 m) and *C.* (*Scarburgiceras*) at 3437 ft (1047·6 m) indicating the *cordatum* Zone and possibly the *costicardia* Subzone.

At 3441 ft 9 in (1049 m) *Cardioceras* (*S.*) *alphacordatum* Spath occurs indicating the *mariae* Zone (*praecordatum* Subzone) with *C.* (*S.*) *praecordatum* Douvillé or *C. bukowskii* Maire at 3444 ft (1049·7 m), *C.* (*S.*) *praecordatum* at 3450 ft (1051·6 m) and *Peltoceras* (*Parawedekindia*) cf. *arduennense* (d'Orbigny) at 3446 and 3449 ft (1050·3, 1051·3 m). No cores were taken in this part of Ashdown No. 2.

Only a small part of the Lower Oxford Clay was cored in Ashdown No. 1, the dark brown bituminous mudstones yielding the long-ranging genus *Hecticoceras* at 3674 ft 10 in (1120·1 m). *Nucula* and *Dicroloma* also occur. Rather more material is available from Ashdown No. 2 as the lowest 30 ft (9·1 m) of the Oxford Clay were cored. *Hecticoceras* was found at 3621 ft (1103·7 m) immediately above *Kosmoceras* (*Spinikosmokeras*) *pollux* (Reinecke) at 3622 ft (1104 m) and *K.* (*Gulielmiceras*) *gulielmi posterior* (Brinkmann) at 3628 ft 6 in (1106 m), *K.* (*Zugokosmokeras*) cf. *obductum* (S. Buckman) at 3634 ft 9 in (1107·9 m) and *K.* (*S.*) *castor* (Reinecke) at 3638 ft (1108·9 m) indicating the *coronatum* Zone with a base at about 3638 ft (1108·9 m).

Kosmoceras (*Gulielmites*) *jason* (Reinecke) at 3639 ft 6 in (1109·3 m) indicates the *jason* Zone and Subzone with further specimens at 3643 ft 8 in, 3646 ft 2 in and 3649 ft (1110·6, 1111·4, 1112·2 m). *Kosmoceras* (*Gulielmiceras*) *gulielmi* (J. Sowerby) occurs at 3645 ft 2 in and 3649 ft 6 in (1111, 1112 4 m); *Choffatia* (*Elatmites*) cf. *submutata* (Nikitin) occurs at 3646 ft 3 in (1111·4 m). The *medea* Subzone is indicated at 3651 to 3652 ft (1112·8–1113·1 m) by *K.* (*Gulielmites*) cf. *medea* Callomon and *K.* cf. *gulielmi*. These bituminous mudstones overlie muddy sandstones of the Kellaways Beds.

Kellaways Beds. Ashdown No. 1, 3721 to about 3752 ft (1134·2–1143·6 m)—about 31 ft (9·5 m).
              Ashdown No. 2, 3652 to about 3682 ft (1113·1–1122·3 m)—about 30 ft (9·1 m).

About 30 ft (9·1 m) of grey muddy sandstones occur in both boreholes. They are generally bioturbated and contain numerous burrows. *Liostrea* cf. *alimena* (d'Orbigny) is abundant though poorly preserved; it occurs both scattered and in groups in the sandstone. *Camptonectes, Chlamys* and *Thracia* were also noted in Ashdown No. 1. *Myophorella* cf. *scarburgensis* (Lycett) at 3679 and 3680 ft (1121·4, 1121·7 m), *Oxytoma expansum* (Phillips) at 3653 ft (1113·4 m) and *Trigonia elongata* J. de C. Sowerby at 3679 ft (1121·4 m) occur in Ashdown No. 2.

Cornbrash and Ashdown No. 1, about 3752 to 3763 ft (1143·6–1147 m)—?11 ft (3·4 m);
Forest Marble. 3763 to 3801 ft (1147–1158·5 m)—38 ft (11·6 m).
              Ashdown No. 2, about 3682 to 3693 ft (1122·3–1125·6 m)—?11 ft (3·4 m); about 3693 to 3715 ft (1125·6–1132·3 m)—22 ft (6·7 m).

Below the Kellaways Beds in both boreholes occur grey mudstones and silty limestones whose exact age is difficult to determine. In Ashdown No. 1 the limestones contain scattered ooliths and a fauna with echinoderm fragments, *Astarte* ?, *Posidonia* ?, *Camptonectes* ? and *Protocardia*. In Ashdown No. 2 the muddy siltstones become more calcareous towards their base at 3693 ft (1125·6 m) but at 3691 to 3692 ft (1125–1125·3 m) contain *Modiolus* cf. *imbricatus* J. Sowerby, *Liostrea* ? and *Placunopsis* ? which are more suggestive of the Forest Marble. These beds are doubtfully placed in the Cornbrash.

They rest on dark grey fine-grained limestones and oolitic limestones with abundant echinoderm and shell fragments. In Ashdown No. 2 a calcareous sandstone occurs at the base of these beds whilst in Ashdown No. 1 the limestone rests on a slight erosion surface. In Ashdown No. 1 the molluscan fauna included *Camptonectes, Chlamys, Liostrea, Lopha* ?, *Modiolus, Placunopsis socialis* ? Morris and Lycett and *Amberleya*. Brachiopods include *Kallirhynchia* at about 3781 ft (1152·5 m) and *Digonella digona* (J. Sowerby) at 3789 ft (1154·9 m) with *Digonella* ? between 3786 and 3790 ft (1154–1155·2 m). No specimens are available from Ashdown No. 2 between 3695 and 4759 ft (1126·2–1450·5 m).

Great Oolite. Ashdown No. 1, 3801 to 3944 ft (1158·5–1202·1 m)—143 ft (43·6 m).
              Ashdown No. 2, 3715 to 3873 ft (1132·3–1180·5 m)—158 ft (48·2 m).

Only a very small fauna was recovered from these limestones and oolitic limestones of which the top 30 ft (9·1 m) were cored in Ashdown No. 1. This included echinoderm fragments, *Lopha* cf. *gregarea* (J. Sowerby) and *Modiolus?*; the sequence is placed in the Great Oolite on lithological grounds. The base is taken at a thin non-oolitic limestone in Ashdown No. 1 and at the base of some muddy grey limestones in Ashdown No. 2.

Fuller's Earth. Ashdown No. 1, 3944 to 4034 ft (1202·1–1229·6 m)—90 ft (27·4 m).
Ashdown No. 2, 3873 to 3957 ft (1180·5–1206·1 m)—84 ft (25·6 m).

Clays occur between the oolitic limestones assigned to the Great Oolite above and Inferior Oolite below. In both cases they contain a sequence of calcareous beds, in the middle, which might be equivalent to the Fuller's Earth Rock. The clays above the calcareous beds in Ashdown No. 1 contain a typical but poorly diagnostic bivalve fauna suggesting the Bathonian—*Arcomya? Anisocardia truncata?* (Morris), *Anisocardia?*, *Entolium corneolum* (Phillips), *Exogyra?*, *Inoperna plicata* (J. Sowerby), *Liostrea?*, *Myophorella?*, *Placunopsis socialis*, *Pseudolimea*, and a malacostracous crustacean claw.

The calcareous beds contain echinoderm, rhynchonellid and terebratulid fragments, also a few species of brachiopods including *Rhynchonelloidella wattonensis* Muir-Wood at 3993 ft (1217·1 m), *Wattonithyris* at 3999 ft (1218·9 m) and *Ornithella?* at 3994 ft (1217·4 m). The bivalves include *Chlamys?*, *Entolium corneolum*, *I. plicata*, *Liostrea?* and *Pseudolimea*.

The lower argillaceous group also contains echinoderm and terebratulid fragments and abundant ostracods, a *Lingula* occurred between 4001 to 4014 ft (1219·5–1223·5 m) and the scanty bivalve fauna includes *Camptonectes?*, *Inoceramus*, *Meleagrinella*, *Nuculana?* and *Plicatula?*

A similar lithological sequence occurs in Ashdown No. 2 but no cores were taken.

Inferior Oolite. Ashdown No. 1, 4034 to about 4445 ft (1229·6–1354·8 m)—about 411 ft (125·3 m).
Ashdown No. 2, 3957 to about 4336 ft (1206·1–1321·6 m)—about 379 ft (115·5 m).

Only two short cores were taken in this group in Ashdown No. 1 and no core in Ashdown No. 2 so the faunal evidence is very limited. The boundaries are taken around the records of oolitic and similar limestones. *Gryphaea?* occurred between 4034 to 4054 ft (1229·6–1235·6 m) and *Camptonectes* between 4100 to 4116 ft (1249·7–1254·5 m).

Lias. Ashdown No. 1, about 4445 to 4538 ft (1354·8–1383·2 m)—93 ft (28·3 m) penetrated.
Ashdown No. 2, about 4336 to about 5628 ft (1321·6–1715·4 m)—1292 ft (393·8 m).

There is a lithological similarity between the oolitic ironstones recorded in Ashdown No. 2 between 4336 and 4378 ft (1321·6–1334·4 m) and the limonitic oolites containing Upper Lias ammonites in the Penshurst (1938) Borehole, and these beds are placed in the Upper Lias as are the beds in Ashdown No. 1 occurring between 4445 and 4515 ft (1354·8–1376·2 m). As in Penshurst there is some lithological variety in these beds, for calcareous siltstones and sandy limestones are also present. In Ashdown No. 1 a thin (4 in (102 mm)) conglomeratic bed occurs at 4515 ft (1762·2 m) consisting of a shelly limestone with irregular pebbles of limonitic siltstone resting fairly sharply on the underlying grey cross-bedded micaceous sandy siltstone which continues to the base of the hole at 4538 ft (1383·2 m).

Dr. Chaloner has examined this sequence for spores and reports:

"Two horizons in the fine-grained current bedded muddy siltstones between 4515 ft 4 in and 4538 ft (1376·3–1383·2 m) have yielded spore material. The sample from 4516 ft (1376·5 m) contains a few specimens of *Classopollis torosus* (Reissinger) Balme, that from 4535 ft (1382·3 m) yielded *C. torosus* and *Deltoidospora*. These are ubiquitous spores throughout the Jurassic."

Dr. J. R. Hawkes reports that the Upper Lias specimen E 38012[1] (CE 1061) from a depth of 4515 ft (1376·2 m) in Ashdown No. 1 is a shelly, limonitic, chamosite-oolite limestone. Shell fragments and chamosite ooliths, false ooliths and flakes are set in a calcite matrix which also contains minor detrital quartz. The outer parts of chamositic components are extensively limonitized (now composed chiefly of ?goethite and kaolinite). Specimen E 38013 (CE 1062) from a depth of 4515 ft (1376·2 m) shows a transition from the above lithology to that of a chamosite-bearing, micaceous, calcareous siltstone. Most of the sample is siltstone, but it contains pockets composed of calcite shell debris and chamosite ooliths cemented by calcite. Limonitic alteration of the chamosite is a minor feature.

The remaining Liassic specimens from the borehole (E 38014–21, CE 1063–70; depths ranging from 4516 ft down to 4535 ft (1376·5–1382·3 m)) are micaceous calcareous siltstones. Each bears small quantities (probably less than 5 per cent) of chamosite in the form of small flakes and rounded aggregates of clay-grade material. Partial alteration of the chamosite to limonite is generally evident. Accessory opaque oxide occurs in most of the specimens and two in particular (E 38017–8, CE 1066–7) contain abundant recrystallized pieces of calcite shell debris.

The term chamosite is used in the sense that the mineral in question closely resembles chamosite in being pale green and of clay mineral habit. X-ray identification of the species has not been carried out.

None of the lithological features noted in specimens E 38014–21 are indicative of the Upper as opposed to the Lower Lias. However, specimen E 38013 suggests transition of the oolite facies into the siltstone lithology without the intervention of a non-sequence. The presence of chamosite in the siltstones could be taken to indicate that this material represents the lower part of a sedimentation cycle like those Taylor (1951, pp. 78–9) described in the Upper Lias of the nearby Penshurst (1938) Borehole (Dines and others 1969, pp. 22–3). Here calcareous siltstones pass gradually upwards into calcareous iron-rich oolites. If this is the case, the Ashdown siltstones between 4516 and 4535 ft (1376·5–1382·3 m) are almost certainly of Upper Lias age.

In Ashdown No. 2 the thick Lias sequence was only cored between about 4757 and 4762 ft (1449·9–1451·5 m), 5270 to 5300 ft (1606·3–1615·4 m) and near the base at 5533 to 5570 ft (1686·5–1697·7 m) so that macrofaunal evidence is again very limited. The highest core, in micaceous silty calcareous mudstone, yielded *Astarte*, *Chlamys*, *Entolium?*, *Oxytoma inequivalve* (J. Sowerby), *Tancredia lucida?* (Terquem) and *Unicardium cardioides?* (Phillips). The record of calcareous silty mudstones and sandy limestones with chamosite ooliths suggests that this horizon may correspond to the lower group of silty ironstones and limestones with ferruginous ooliths found between 4040 and 4060 ft (1231·4–1237·5 m) in the Penshurst (1938) Borehole and which were thought to be Middle Lias. A similar group of ferruginous beds was identified in the Warlingham Borehole (Worssam and Ivimey-Cook 1971) as belonging to the *spinatum* Zone on brachiopod evidence and it overlies beds with ammonites of the *margaritatus* Zone. The fauna from Ashdown No. 2 is quite consistent with an horizon in the Middle Lias.

The next specimens, from between 5274 ft 6 in and 5290 ft 9 in (1607·7–1612·6 m) yielded *Cincta?*, *Calcirhynchia* or *Squamirhynchia*, *Spiriferina?*, *Camptonectes*, *Cardinia attenuata?* (Stutchbury), *Gryphaea* and *Pseudopecten*. This assemblage suggests a horizon in the Sinemurian or Lower Pliensbachian.

---

[1]Numbers prefixed by E refer to specimens in the English Sliced Rock Collection of the Institute of Geological Sciences.

A few further specimens were obtained from the cores taken below 5533 ft (1686·5 m). *Pteria longiaxis* (J. Buckman) at 5533 ft (1686·5 m) and *Modiolus sp.* between 5559 and 5620 ft (1694·4–1713 m) occur with bivalve fragments and abundant ostracods and suggest that these beds are Lower Lias. Dr. Chaloner has identified the Rhaeto–Jurassic spore *Classopollis torosus* from mudstone at 5620 ft (1713 m). Below, at 5624 ft (1714·2 m) are hard greyish green and greyish brown calcilutites with pyritic flecks which rest on a red mudstone at 5628 ft (1715·4 m). The age of these beds is not known, as similar limestones occur both in the Lower Lias and in the Upper Triassic. In view of the great change in lithology immediately below they are doubtfully included in the Jurassic.

Beds of Unknown Age. Ashdown No. 2, 5628 to 5709 ft (1715·4–1740·1 m)—81 ft (24·7 m) base of borehole

Below 2 ft (0·6 m) of red mudstone occur 22 ft (6·7 m) of brecciated mudstone resting on brownish red silty and calcareous mudstones and sandstones. No evidence has been found for the age of these beds. Comparison with the Henfield Borehole (Chaloner 1962) suggests that they also might be very tentatively assigned to the Trias.

H.C.I.-C.

ADDENDUM

Since this memoir went to press, the deep Cowden No. 1 Borehole [TQ 4668 4278] has been completed in the autumn of 1971; it was sited on the north-eastern flank of the Gilridge Anticline on the adjacent Sevenoaks (287) Sheet (Dines and others 1969, p. 9). First reports, kindly made available by Ball and Collins (Oil and Gas) Limited, BP Petroleum Development Limited and Gas Council (Exploration) Limited, indicate the presence of 5200 ft (1524 m) of Jurassic strata, the greatest amount so far proved in the north and central part of the Wealden Basin. There is a possibility that this thickness has been augmented by fault duplication of part of the sequence.

Chapter IV

JURASSIC: PURBECK BEDS

THE OLDEST rocks seen at the surface in south-eastern England are the Purbeck Beds present in the core of the Wealden anticlinorium. The three major inliers of Purbeck Beds in East Sussex have been studied in detail in conjunction with the survey of this sheet and those it adjoins. The results of this work have recently been published (Anderson and Bazley 1971). Only a résumé of the conclusions of that work will be repeated here, as with the work of Howitt (1964) these strata are considered from most general aspects to have been adequately described.

On this sheet the Purbeck Beds crop out along the south-east margin in part of the Broadoak–High Wood (Brightling) Inlier. Here they are about 480 ft (146·3 m) thick but towards Crowborough in the north-west they thicken to about 620 ft (189 m) (Ashdown Nos. 1 and 2 boreholes). The oldest rocks at the surface are the Blues Limestones (Fig. 3). Below these, in the lowest 50 ft (15·2 m) of the Purbeck Beds, are gypsum and anhydrite deposits that are of considerable economic importance (p. 116).

By using the ostracod assemblages a number of divisions of the Purbeck have been established (Anderson and Bazley 1971). The Blues Limestones, about 40 ft (12·2 m) thick, are at the top of the Lower Purbeck and consist of shelly calcareous mudstones and highly crystalline limestones. These are exposed in Milkhurst Wood stream [6260 2197], east of Tottingworth Park; upstream many of the younger beds of the Purbeck can be seen (Bazley and Bristow 1965). In all the tributaries flowing down from the height of Burwash Common to the River Dudwell there are exposures of the Middle and Upper Purbeck.

Immediately above the Blues Limestones calcareous mudstones dominate the succession and about 40 ft (12·2 m) higher, in the Middle Purbeck, there is a series of limestones and shales, about 5 ft (1·5 m) thick, that have been correlated with the Cinder Beds horizon of Dorset. This horizon commonly includes *Liostrea distorta* (J. de C. Sowerby). Above are over 100 ft (30·5 m) of rhythmically deposited argillaceous limestones, shales and mudstones, silts and sandstones. Towards the top of the Middle Purbeck the sandstone member of each rhythm becomes better developed and is frequently exposed in the streams. These are succeeded by the Upper Purbeck Greys Limestones group of dark grey mudstones, shales and limestones. Near the base clay-ironstones are developed in a predominantly mudstone sequence but above come extremely shelly limestones and shales (Plate IIIA) that include some distinctive leathery 'paper' shales. The upper part of the Upper Purbeck is predominantly clay with subordinate silts, sandstones and nodular clay ironstones.

The macrofauna of the very fossiliferous Purbeck Beds is composed almost entirely of bivalves and gastropods. These commonly include *Neomiodon sp.*, *Corbula sp.*, *Unio sp.* and *Viviparus sp.* Fish bones and scales are present through-

25

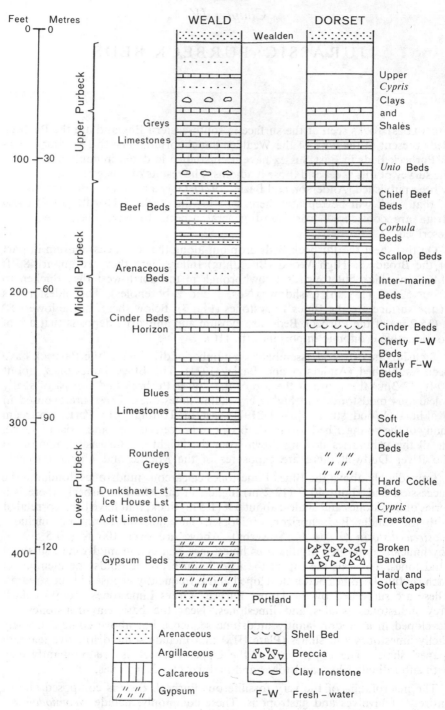

FIG. 3. *Generalized lithological sequence and main divisions of the Purbeck in the Brightling area, compared with Dorset*

(A 10315)

A. Upper Purbeck Beds (Greys) in Milkhurst Wood, Burwash

PLATE III

B. Massive top Ashdown Beds at Sandhurst Farm, near Lamberhurst

(A 10320)

out and plant remains are common. The lack of a truly marine fauna is interesting and it is suggested that the Purbeck Beds were deposited in a lagoon with restricted circulation of water and variable conditions of salinity. Throughout the Purbeck sedimentation was rhythmic, the rhythms probably controlled by minor epeirogenic oscillations. The sedimentary rhythms do not always correspond with faunal rhythms expressed in changes of the ostracod fauna from sequences dominated by forms tolerant of high salinity to others made up mainly of relatively fresh-water species. These faunal rhythms may have been controlled by climatic changes associated with the 21 000-year cycle of the precession of the equinoxes (Anderson and Bazley 1971).

The Cinder Beds horizon marks one of several major episodes of increased salinity recognized in Sussex (Anderson and Bazley 1971). It has been proposed (Casey 1963) that their lower limit should be taken as the base of the Cretaceous and that the Cinder Bed and Purbeck Beds lying above it be called the Durlston Beds. Norris (1969), however, whilst accepting that the Cinder Bed of Dorset marks the base of the Cretaceous, thought that this correlated with the gypsiferous series at Mountfield and that the Cinder Beds in Sussex correlated with one of the Chief Beef Beds at Durlston. Norris's microfloral correlation (1969, p. 614) places the entire Sussex Purbeck within the Cretaceous System. The Cinder Beds horizon is not one that is lithologically significant at the surface in the Sussex area; both their lithology and fauna make the Upper, Middle and Lower Purbeck divisions easily determinable and for this reason these divisions have been necessarily retained.

Hughes and Moody-Stuart (1969) found miospores that characterize the lower part of the 'Fairlight Clay' (see p. 36), as exposed on the coast, to occur between 1915 ft and 1968 ft 10 in (583·7–600·1 m) in the Warlingham Borehole, i.e. in beds that lithostratigraphically are Purbeck Beds (Greys Limestones with 29 ft 10 in (9·1 m) of overlying mudstones). The implied correlation between the 'Fairlight Clay' and the upper part of the Purbeck Beds of Warlingham is accepted by Norris (1969) but not by Worssam and Ivimey-Cook (1971).

It was found that remains of old bell-pits are a feature of the Purbeck outcrop between Perch Hill and Tottingworth Park. They were dug at several horizons for limestone and clay-ironstone and leave typically uneven land (see Gould *in* Topley 1875, pp. 384–6). In the Upper Purbeck there are occasional pits for clay-ironstone but the vast majority were dug in the Greys Limestones and Blues Limestones groups. Most of the limestone produced was used as agricultural lime (p. 114).   R.A.B.

Chapter V

CRETACEOUS

GENERAL ACCOUNT

WEALDEN: HASTINGS BEDS

THE SUBDIVISIONS OF THE HASTINGS BEDS

THE TERM Hastings Sands (now generally referred to as the Hastings Beds) was first used by Fitton (1824) to describe the mixed series of ferruginous sandstones, silts and shales lying between the Weald Clay and the shale–limestone sequence of the Purbeck. Mantell made two attempts to subdivide the group on a lithological basis: in 1822 into the Tilgate Beds and the Iron Sand and in 1827 into the Tilgate Beds, Worth Sandstone and Horsted Sand. Both attempts were unsuccessful because neither scheme was based on geological mapping and consequently both contained numerous errors of correlation.

In 1861 Drew proposed the now accepted subdivisions of the Hastings Beds as a result of his work on Old Series Geological Sheet 6. He suggested that in the northern part of the High Weald Mantell's 1827 scheme was unworkable and should be replaced by the formations Ashdown Sand, Wadhurst Clay and Tunbridge Wells Sand; in support of this suggestion he outlined areas of outcrop for the proposed formations. He further proposed the name Grinstead Clay for the persistent bed of clay which separates the Lower and Upper Tunbridge Wells Sand in the East Grinstead area. Later, as a result of the same survey, Gould (*in* Topley 1875, p. 7) introduced Fairlight Clays to describe the argillaceous beds in the lower part of the Ashdown Sand at Hastings, and Bristow (*in* Topley 1875, pp. 4 and 91) the term Cuckfield Clay for a clay bed in the Upper Tunbridge Wells Sand at Cuckfield.

More recently, the terms Middle Tunbridge Wells Sand (Milner 1923a; Reeves 1949), Balcombe Clay (Reeves 1949), Lingfield Beds (Edmunds 1935) and Lower Tunbridge Wells Clay (Michaelis 1968) have been used to describe local facies of the Tunbridge Wells Sand. Names have also been introduced to describe sandstone beds of local importance within the Wadhurst Clay (Hog Hill Sand; Allen 1947), within the Grinstead Clay (Cuckfield Stone; Gallois 1963) and within the Lower Tunbridge Wells Sand (Ardingly Sandstone; Gallois 1965a).

At the present time any subdivision of the Hastings Beds must be based on lithological characters which can be recognized in the field, since these are the characters which have been used in making the present map and the New Series maps of the adjacent areas. It follows naturally from this that the maps themselves form part of the definition of the formations described below.

Some of the most important recent work on the Hastings Beds has been the intensive study of the sedimentology by Allen (1949a–1967b). In addition to

28

providing a very clear picture of the conditions of deposition of these sediments (see p. 38) Allen has demonstrated the presence of several marker beds within the sequence and has shown them to have a remarkable lateral continuity. The two most important of these, from the point of view of defining the formation boundaries, are the Top Ashdown Pebble Bed and the Top Lower Tunbridge Wells Pebble Bed. A generalized section of the Hastings Beds succession of the western High Weald, where the sequence is most completely subdivided, showing the position of these and other marker beds is given in Fig. 4.

When one considers the deltaic origin of the Hastings Beds sediments one might expect the formations described below to be diachronous, but until such time as a detailed zonal scheme has been worked out for the full sequence over the whole of its outcrop one can merely speculate on this point. It is to be hoped that the detailed ostracod zonal scheme recently proposed by Anderson and others (1967, p. 188) for the Wadhurst Clay will eventually be extended to cover the remainder of the Hastings Beds which, as yet, have only an outline zonal classification. However, lack of suitable fossil material in the arenaceous formations will always make palaeontological classification difficult.

The designation of type sections in the Hastings Beds is impracticable. Complete natural sections are unknown for any of the formations and artificial sections are subject to change or rapidly become degraded. Cored boreholes provide the most continuous sections available to us at the present time. For the above reasons several reference sections are given for each formation and, wherever possible, one of these is within the area described in the original definition. The original type areas of Drew (1861) all fall within the confines of the Tunbridge Wells map. The use of the terms 'group', 'formation' and 'member' follows that defined by the American Commission on Stratigraphic Nomenclature (Anon. 1961).

DETAILED DESCRIPTIONS OF THE HASTINGS BEDS GROUP

"It may be advantageous therefore . . . to adopt one [a name] from a place where, as at Hastings, the strata are well developed and conspicuous . . . The Hastings Sands in the Isle of Wight may be described as consisting of a series of beds of sand . . . alternating with beds of clay, much mixed with sand, of a greenish or reddish hue, or of a mottled and variegated appearance" (Fitton 1824, p. 377). Accompanying sections show the Hastings Sand to lie between the Purbeck Beds and the Weald Clay. The strata seen in the cliffs at Hastings (mostly Ashdown Beds) are too arenaceous to be representative of the Group as a whole and the name Hastings Beds, instead of Hastings Sands has been in common usage since 1875 (Topley, p. 2).

The following formations are described in ascending stratigraphical order:

1. Ashdown Beds (Formation)

"The lower sands make very high ground, are spread over a large area, and show a very great thickness on Ashdown Forest . . . no name, I think can be better for these than Ashdown Sand" (Drew 1861, p. 283). Because of the confusion which has arisen concerning the relationship of the Fairlight Clays to the Ashdown Sand (see below), despite the clarity of Gould's definition (*in* Topley 1875, p. 7) of the former name, the term Ashdown Beds (Dines and others 1969, p. 29) has been used to describe this formation.

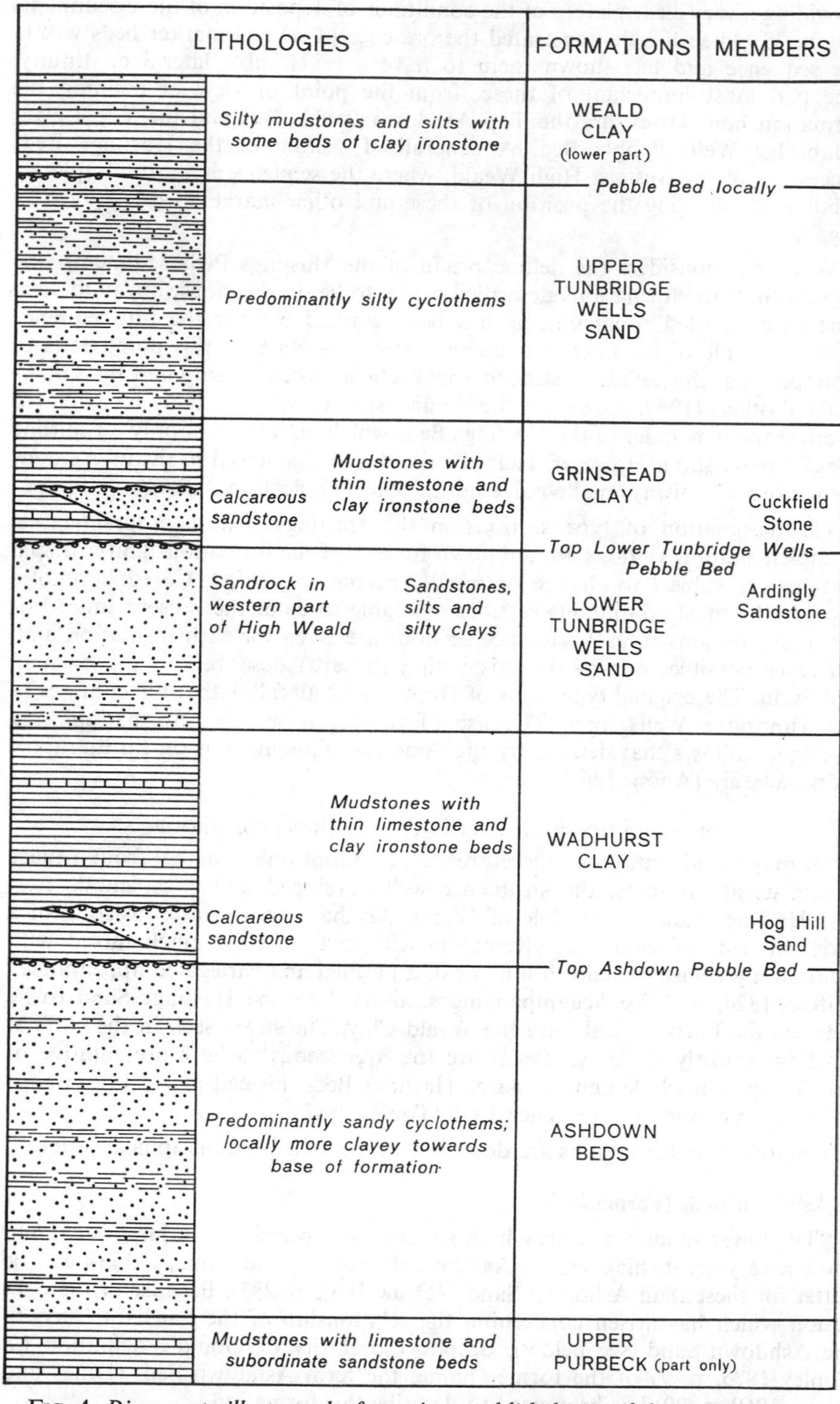

LITHOLOGIES	FORMATIONS	MEMBERS
Silty mudstones and silts with some beds of clay ironstone	WEALD CLAY (lower part)	
	Pebble Bed locally	
Predominantly silty cyclothems	UPPER TUNBRIDGE WELLS SAND	
Calcareous sandstone — Mudstones with thin limestone and clay ironstone beds	GRINSTEAD CLAY	Cuckfield Stone
	Top Lower Tunbridge Wells Pebble Bed	
Sandrock in western part of High Weald — Sandstones, silts and silty clays	LOWER TUNBRIDGE WELLS SAND	Ardingly Sandstone
Mudstones with thin limestone and clay ironstone beds	WADHURST CLAY	
Calcareous sandstone		Hog Hill Sand
	Top Ashdown Pebble Bed	
Predominantly sandy cyclothems; locally more clayey towards base of formation·	ASHDOWN BEDS	
Mudstones with limestone and subordinate sandstone beds	UPPER PURBECK (part only)	

FIG. 4. *Diagram to illustrate the formations and lithologies of the Hastings Beds*

The Ashdown Beds are a predominantly arenaceous formation consisting of fine-grained sandstones and siltstones with subordinate amounts of shale and mudstone. The junction with the underlying Purbeck Beds is marked by a lithological change from silty sandstone or siltstone to the soft mudstones of the Upper Purbeck (see p. 25; Holmes *in discussion of* Howitt 1964, p. 111; Anderson and Bazley 1971). The top of the Ashdown Beds is marked by the Top Ashdown Pebble Bed, the transgressive base of the overlying Wadhurst Clay; this boundary is commonly accompanied by an abrupt change from massive or thickly bedded sandstones to mudstones and shales. In Ashdown Forest the upper part of the Ashdown Beds, the only part which crops out, is made up of a number of predominantly sandy cyclothems (see p. 51). Elsewhere, evidence from outcrops and boreholes shows that the cyclothems in the lower part of the formation are locally more clayey. Individual clay beds have been mapped at several levels within the formation but none of these has yet been found to be either sufficiently distinctive or laterally continuous to justify being named. The term Fairlight Clays is discussed below under proposed deletions.

The estimated thickness of the Ashdown Beds is 750 ft (228·6 m) at Crowborough (see p. 39).

Reference sections

Ashdown Forest area; no good sections are present but small exposures occur along many of the stream courses in the area (see p. 51 for details).

Hastings Cliffs; almost continuous sections from Hastings to Pett Level; see Topley (1875, pp. 45–9, 54–5) and Allen (1962, pp. 219–25).

Jarvis Brook Brickworks, Crowborough [531 297]; 50 ft (15·2 m) of strata in the middle of the formation; see also p. 56.

Cuckfield No. 1 Borehole [2961 2731]. I. G. S. Specimen Nos. Bz 3244–3417; (see Lake and Thurrell, *in press*).

Old quarry at Elphicks, Goudhurst [699 381]; junction of Ashdown Beds and Wadhurst Clay with Top Ashdown Pebble Bed present; see also Shephard-Thorn and others (1966, p. 28) and Allen (1949, p. 268) for other sections at this level.

Old quarry at High Hurstwood, Buxted [499 271]; junction of Ashdown Beds and Wadhurst Clay with Top Ashdown Pebble Bed present; see Plate VA and p. 53.

2. Wadhurst Clay (Formation)

" . . . near the village of Wadhurst it [the clay] spreads over some tolerably high ground, which is indented by a series of valleys that expose a great thickness of it, and at least reach down to the sands below . . . I have on these accounts thought it well to take the name of Wadhurst Clay for the second stratum" (Drew 1861, p. 283).

The Wadhurst Clay is an argillaceous formation consisting of soft dark grey shales and mudstones with subordinate beds of siltstone, sandstone, shelly limestone and clay ironstone (see Anderson and others 1967). At the base is the Top Ashdown Pebble Bed: the top is marked by a change from shales to the siltstones and fine-grained silty sandstones of the basal Tunbridge Wells Sand. The top few feet commonly weather to a stiff red clay. The Wadhurst Clay shows little lithological variation within the area of its outcrop: its thickness varies from 100 ft (30·5 m) at Rye to about 230 ft (70·1 m) in the East Grinstead area.

2(a). *Hog Hill Sand* (Member). A number of sandstone bodies having varying lateral extents occur within the Wadhurst Clay and some uncertainty attends the correlation of these bodies from place to place. They are particularly prominent in the Hastings district, where Topley (1875, p. 62) noted "The Wadhurst Clay near Icklesham has a good deal of sand and sandstone in it . . . there are perhaps 20 ft [6·1 m] of it, with 25 or 30 ft [7·6 or 9·1 m] of clay between it and the Ashdown Sand". This sandstone has subsequently been named the Hog Hill Sand (Allen 1959, p. 296) from Hog Hill [887 160], Icklesham, and has been regarded as the lowest sandstone in the Wadhurst Clay, occurring some 15 to 30 ft (4·6–9·1 m) above the base of the formation (Allen 1959, p. 297). Recent remapping of the Hog Hill district indicates that the sandstone capping the hill there may be the second sandstone above the base of the Wadhurst Clay and may be of relatively local extent.

Both the top and bottom of the Hog Hill Sand are marked by an abrupt change from sandstone to shales, the upper boundary being further marked by a pebble bed rich in bone fragments (Telham Bone Bed, Allen 1949b, p. 279). Throughout most of its outcrop the sandstone varies from 0 to 20 ft (6·1 m) in thickness (Shephard-Thorn and others 1966, pp. 44, 60–62; Allen 1959, p. 297), but locally, where the sandstone cuts down nearly to the base of the Wadhurst Clay it exceeds this thickness (Allen 1959, p. 297).

Reference sections

Wadhurst Park Boreholes [632 291]. I. G. S. Specimen Nos. Bw 7085–7523; see Anderson and others (1967).

Sharpthorne Brickworks, West Hoathly [374 329]; basal 30 ft (9·1 m) of the formation; junction with Ashdown Beds sometimes visible in works below pit.

Freshfield Lane Brickworks [382 266], lower workings; top 50 ft (15·2 m) of formation including red clays and High Brooms Soil Bed; disturbed by valley bulging; see also Gallois (1964, pp. 361–2).

High Brooms Brickworks, Southborough [594 418]; top 65 ft (19·8 m) of formation including red clays, High Brooms Soil Bed and junction with Lower Tunbridge Wells Sand; see also Dines and others (1969, p. 41).

Cuckfield No. 1 Borehole [2961 2731]. I. G. S. Specimen Nos. Bz 2908–3243; see Lake and Thurrell (*in press*).

Quarry Hill Brickworks, Tonbridge [585 450]; 30 ft (9·1 m) of beds in the middle part of the formation, some 70 ft (21·3 m) above the Ashdown Beds; faulted against Tunbridge Wells Sand; see also Dines and others (1969, pp. 11, 40, 44).

3–5. Tunbridge Wells Sand (Sub-group)

" . . . about Tunbridge [Wells] the upper sands occupy a great space; they form nearly all the hills near, and their top bed makes those conspicuous rocks I have spoken of. I think, therefore, their name is well taken from that place" (Drew 1861, p. 283).

In the western part of the High Weald the Tunbridge Wells Sand is divided into three parts, a lower and upper sandy formation separated by the Grinstead Clay, but in the eastern part of the Hastings Beds outcrop it is not possible to make this subdivision. The junction with the Weald Clay is broken by faulting or obscured by Recent deposits in the eastern tract. A maximum of 250 ft (76·2 m) of strata has been proved for the undivided sequence.

In the western High Weald the following three formations are recognized.

3. Lower Tunbridge Wells Sand (Formation)

The Lower Tunbridge Wells Sand is a predominantly arenaceous formation and consists of fine-grained sandstones with subordinate, but appreciable, amounts of siltstone and thin beds of silty clay. Throughout much of its outcrop the formation can be divided into a lower part consisting of interbedded silt-stones and fine-grained silty sandstones, overlain by a massive sandstone (see below). The base of the formation is described above: the top is usually marked by the Top Lower Tunbridge Wells Pebble Bed (see Allen 1959, p. 309), the transgressive base of the Grinstead Clay. This upper boundary is invariably accompanied by a sharp change from massive sandstone to shales and is analogous to the Ashdown Beds–Wadhurst Clay junction.

The Lower Tunbridge Wells Sand is about 130 ft (39·6 m) thick at Tunbridge Wells, 110 ft (33·5 m) at East Grinstead and 90 ft (27·4 m) at Cuckfield.

3(a). *Ardingly Sandstone* (Member). The massive sandstone referred to above as occupying the upper part of the Lower Tunbridge Wells Sand forms many conspicuous outcrops in the Ardingly to Balcombe area and has been termed the Ardingly Sandstone (Gallois 1965a, p. 47). The bulk of the sandstone consists of massive, cross-bedded, fine-grained, quartzose, poorly cemented sandstone (sandrock). The upper and lower few feet of the Ardingly Sand-stone are lithologically more variable than the remainder and consist locally of silty, ferruginously cemented, thickly and flaggy bedded sandstone (see Dines and others 1969, pl. IIIA). The top of the sandstone is taken at the Top Lower Tunbridge Wells Pebble Bed; the base, where seen, is marked by an abrupt change from thickly bedded or massive sandstones to thinly interbedded siltstones and silty sandstones of the lower part of the Lower Tunbridge Wells Sand.

In the Tunbridge Wells area this sandstone is commonly medium grained and contains stringers and lenses of small pebbles. To the south-west it becomes fine grained and thickly bedded, in places losing its distinctive sandrock lithology. Eastwards from Tunbridge Wells and from Uckfield it passes into a sequence of interbedded sandstones and siltstones, where it can no longer be distinguished.

The Ardingly Sandstone forms many conspicuous natural outcrops, particu-larly in the Ardingly, Balcombe, East Grinstead, Eridge, Tunbridge Wells and Uckfield areas.

The outcrops referred to by Drew (see above) as being the top bed of the Tunbridge Wells Sand are within the Lower Tunbridge Wells Sand (Buchan 1938, p. 407) and have recently been shown to be the Ardingly Sandstone (Dines and others 1969, pp. 42–4).

This member is about 55 ft (16·8 m) thick at Tunbridge Wells, 60 ft (18·3 m) at East Grinstead and 40 ft (12·2 m) at Cuckfield.

Reference sections

High Brooms Brickworks, Southborough [594 418]; basal 50 ft (15·2 m) of formation and junction with Wadhurst Clay; see also Dines and others (1969, p. 41).

Stream section near Ditton Place, Handcross [2821 2980]; junction of Ardingly Sandstone and lower part of the Lower Tunbridge Wells Sand.

Cuckfield No. 1 Borehole [2961 2731]; I. G. S. Specimen Nos. Bz 2805–2907; see Lake and Thurrell (*in press*).

Philpots Quarry, West Hoathly [3536 3215]; top of Ardingly Sandstone and junction with Grinstead Clay; see also Allen (1959, p. 301; 1962, pp. 236–41).

Natural crags of Ardingly Sandstone are common, most of those following being chosen largely because they mark the limits of the outcrop: Stonehurst, Ardingly [344 317]; Balcombe Mill [317 305]; Chiddingly Wood, West Hoathly [349 321]; Stone Farm Rocks, Saint Hill Green [381 348]; Redleaf House, Penshurst [522 455]; Rusthall Toad Rock, Tunbridge Wells [568 395]; Eridge Rocks [554 357]; Bowles Rocks, Boarshead [542 330]; The Rocks, Uckfield [464 217]. See also pp. 77–91 and Dines and others (1969, pp. 42–4).

4. Grinstead Clay (Formation)

"The bed of clay that appears in the Tunbridge Wells Sand . . . may well be called the Grinstead Clay, from its occurring on the north side of the town of East Grinstead . . . " (Drew 1861, p. 283).

The Grinstead Clay consists of soft mudstones and silty mudstones with subordinate thin beds of siltstone, clay ironstone and shelly limestone. Over part of its outcrop it is divided by a bed of calcareous sandstone, the Cuckfield Stone (see below), into Lower and Upper Grinstead Clay.

The base of the Grinstead Clay is marked by the Top Lower Tunbridge Wells Pebble Bed or by a minor erosion surface and the junction with the Upper Tunbridge Wells Sand by a change from mudstones to siltstones and silty sandstones. This latter junction is analogous to the Wadhurst Clay–Lower Tunbridge Wells Sand junction and the topmost Grinstead Clay commonly weathers to a red clay similar to the top Wadhurst Clay. This reddening extends down in places as far as the Cuckfield Stone, but rarely reaches the Lower Grinstead Clay.

The outcrop of the Grinstead Clay is broken by faulting and extends from Leigh (where it is 20 to 25 ft (6·1–7·6 m) thick), west of Tonbridge, to Uckfield (30 ft (9·1 m)) via East Grinstead (50 to 60 ft (15·2–18·3 m)) and Cuckfield (60 to 80 ft (18·3–24·4 m)).

4(a). *Cuckfield Stone* (Member). "Another sandstone . . . here designated the Cuckfield Stone, occurs in the middle of the Grinstead Clay at Balcombe, Ardingly, West Hoathly and Cuckfield" (Gallois 1963, p. 37). More recent work has shown that the outcrop extends to near Penshurst on the Sevenoaks (287) Sheet and Langton Green on the Tunbridge Wells (303) Sheet. Both the lower and upper limits of this sandstone are lithologically well defined, the latter being locally marked by the Cuckfield Pebble Bed (Gallois 1964, p. 364). The Cuckfield Stone is 5 to 15 ft (1·5–4·6 m) thick over much of its outcrop and consists of flaggy and thickly bedded sandstones with lenses of calcareous sandstone. It is absent north-west of a line running from East Grinstead to Handcross, and generally thickens south-eastwards away from this line. Its maximum thickness of about 25 to 30 ft (7·6–9·1 m) occurs near Cuckfield and near Finch Green [507 415]. In the Chailey area the base of the member cuts down into the underlying mudstones in a manner comparable to that in which the Hog Hill Sand (see above) locally cuts down.

Reference sections

Philpots Quarry, West Hoathly [3536 3215]; complete section of Lower Grinstead Clay and Top Lower Tunbridge Wells Pebble Beds, and a frost-disturbed junction with Cuckfield Stone; see also Allen (1959, p. 301).

Railway cutting 1 mile (1·6 km) N. of Haywards Heath Station [3284 2642]; complete section of Lower Grinstead Clay with a poorly developed Top Lower Tunbridge Wells Pebble Bed and a disturbed junction with Cuckfield Stone.

Freshfield Lane Brickworks, access road [3830 2642]; complete section of Lower Grinstead Clay, Cuckfield Stone and Upper Grinstead Clay with unusually thick development of Cuckfield Stone; Top Lower Tunbridge Wells Pebble Bed and also the junction with Upper Tunbridge Wells Sand can be seen; disturbed by faulting and cambering; see also Gallois (1964, p. 362).

Old quarry near Whitemans Green, Cuckfield [2986 2546] exposes Cuckfield Stone; see also Gallois (1964, p. 363).

Cuckfield No. 1 Borehole [2961 2731]. I. G. S. Specimen Nos. Bz 2697–2804, see Lake and Thurrell (*in press*).

Old quarries [493 377] near Perryhill Farm, north of Withyham still expose several feet of massive, calcareous Cuckfield Stone.

Old quarry near Shovelstrode Lane, East Grinstead [4100 3578]; junction of Grinstead Clay and Ardingly Sandstone with Top Lower Tunbridge Wells Pebble Bed present; see also Milner (1923b, p. 287, pl. 20b) and Bazley and Bristow (1965, p. 315); Allen (1959, p. 304) gives other sections at this level.

Roadside section [5442 4593] 220 yd (201 m) S. of Paul's Farm, Leigh: Grinstead Clay and its junction with Ardingly Sandstone; see also Dines and others (1969, p. 47).

5. Upper Tunbridge Wells Sand (Formation)

The Upper Tunbridge Wells Sand is a rhythmic sequence of fine-grained sandstones, silty sandstones and siltstones with subordinate amounts of soft mudstone. Thin impersistent clay ironstone beds also occur locally. Siltstones occupy the major part of the formation. Over much of the northern outcrop the junction with the overlying Weald Clay is cut out by faulting or obscured by drift deposits. Where the junction has been observed, notably between Crawley and Lingfield and from Horsham to Wivelsfield, it is most conveniently taken as the top of the highest sandstone of the Tunbridge Wells Sand. In the Crawley to Lingfield area a persistent sandstone bed, 3 to 5 ft (0·9–1·5 m) thick, separates the interbedded sandstones, siltstones and mudstones of the Upper Tunbridge Wells Sand from the silty mudstones and siltstones without sandstones of the lower Weald Clay. Similarly between Horsham and Wivelsfield this boundary is marked by a prominent sandstone up to 15 ft (4·6 m) thick. At several localities near Crawley, Horsham and Wivelsfield the sandstone bed is capped by a thin pebble bed similar to those capping the other Hastings Beds arenaceous formations. Faulting makes it difficult to assess the thickness of the formation, but it is estimated to be 250 to 350 ft (76·2–106·7 m) thick over much of its outcrop. The Cuckfield No. 1 Borehole proved a total thickness of 328 ft (100 m) of Upper Tunbridge Wells Sand (*in A. Rep. Inst. geol. Sci. for 1966*, p. 63).

Reference sections

Cuckfield No. 1 Borehole [2961 2731]. I. G. S. Specimen Nos. Bz 2156–2696; see Lake and Thurrell (*in press*).

Freshfield Lane Brickworks, upper workings [386 264]: basal 40 ft (12·2 m) of the formation and junction with Grinstead Clay.

Hundred Acres Wood Brickworks, Turners Hill [331 365]: 20 ft (6·1 m) of strata in the lower part of the formation, about 50 ft (15·2 m) above the Grinstead Clay.

PROPOSED DELETIONS

(*i*) *Fairlight Clays.* " . . . the clays of the Hastings Cliffs will be described as Fairlight Clays, a term employed by Mr. C. Gould when mapping the beds" (Topley 1875, p. 7). Further, "The beds first seen east of Hastings are alternations of clay and sandstone, the former predominating, as is the case throughout the whole series. The junction is not very definite, although as a whole the Fairlight Clays are well distinguished from the Ashdown Sand by the quantity of clay which they contain" (Topley 1875, p. 47). In practice the Fairlight Clays can only be recognized with certainty in the type section at Fairlight Glen.

The Institute of Geological Sciences Fairlight Borehole [859 117] commenced in the basal beds of the Wadhurst Clay. The Ashdown Beds, some 600 ft (182·9 m) thick, comprised an upper 120 ft (36·6 m) of alternating sandstone, silts and mudstones, predominantly with mudstones (some with sphaerosiderite and red mottling), and silts with only subordinate sandstones, below. Inland, recent six-inch mapping for the Hastings (320) Sheet has demonstrated that clays of 'Fairlight' facies are developed throughout the Ashdown Beds; also in some areas Wadhurst Clay had been incorrectly mapped as Fairlight Clays.

In recent years the slightly modified term Fairlight Clay has commonly been used to describe a transitional clay formation between the Ashdown Beds and the Purbeck Beds (see also p. 29). Mapping of the Purbeck inliers on the Tunbridge Wells (303), Tenterden (304), Lewes (319) and Hastings (320) sheets has demonstrated that the lowest 50 ft (15·2 m) of the Ashdown Beds contain discrete, lenticular mappable clay beds, in a silty and sandy sequence.

Owing to the confusion of the terms Fairlight Clays, Fairlight Clay and Fairlight facies it is proposed that the name be dropped. Any clay seams sufficiently distinctive to be mapped are regarded as clay beds within the Ashdown Beds, and if considered to be sufficiently important can be separately named.

(*ii*) *Lower Tunbridge Wells Clay.* "The Lower Tunbridge Wells Sand . . . includes a lenticular clay member. This clay reaches a maximum thickness of

50 ft (15 m) where it outcrops in the valley of the River Ouse north of Cuckfield"
(Michaelis 1969, p. 530). This clay has been shown (Gallois 1970, p. 169) to be
part of the Weald Clay. Other areas of outcrop of this clay given by Michaelis
(1969, p. 531) fall within the Wadhurst Clay, the Grinstead Clay and the Upper
Tunbridge Wells Sand.

(*iii*) *Cuckfield Clay.* "In the neighbourhood of Cuckfield the Upper Tunbridge
Wells Sand is itself divided by a bed of clay—the Cuckfield Clay" (Topley
1875, p. 4). Recent work has shown that the clay referred to is part of the
Grinstead Clay (Gallois 1964, p. 364) and is therefore covered by the existing
nomenclature.

(*iv*) *Balcombe Clay.* This name has been used to describe a clay, up to 30 ft
(9·1 m) thick, in the Upper Tunbridge Wells Sand of the Balcombe area (Reeves
1948, pp. 245, 255). It is said to be some 45 to 90 ft (13·7–27·4 m) above the
Grinstead Clay and tentative correlation is made with the Cuckfield Clay. No
evidence was found of such a clay bed at this level when the area was surveyed,
but since Reeves provides neither a map of its outcrop nor descriptions of
sections within it, it is difficult wholly to deny its existence.

(*v*) *Middle Tunbridge Wells Sand.* This term has been used by Reeves (1948, p.
245) to describe the beds between the Grinstead Clay and the Balcombe Clay
and, by correlation, between the Grinstead Clay and the Cuckfield Clay (1948,
p. 255). The term is unnecessary (see above).

Middle Tunbridge Wells Sand has also been used, but in a different sense, by
Milner (1923a, p. 48) for the Tunbridge Wells area. The beds assigned to this
unit have been mis-correlated and belong with the Ashdown Beds (Bazley and
Bristow 1965, p. 317).

(*vi*) *Lingfield Beds.* No description exists for this term and Edmunds (1935, p.
20) merely shows the name placed in a stratigraphical table between the Upper
Tunbridge Wells Sand and the Weald Clay. In his description of the highest
Upper Tunbridge Wells Sand of the Lingfield area Dines (*in* Dines and Edmunds
1933, p. 30) contrasted the interbedded siltstones and sandstones of that area
with the sandrock of the Tunbridge Wells area, which at that time was thought
to be at an equivalent horizon within the Upper Tunbridge Wells Sand. The
name was therefore probably introduced to describe the Lingfield facies of the
top Upper Tunbridge Wells Sand. Buchan (1938, p. 407) subsequently showed
that the sandrock outcrops were within the Lower Tunbridge Wells Sand and
more recent work (Dines and others 1969, pp. 42–4) has confirmed this. The
lithology of the Upper Tunbridge Wells Sand of the Lingfield area is similar
to that of other areas and the name is therefore unnecessary.

R.W.G., C.R.B., R.A.B., E.R.S.-T.

HASTINGS BEDS

In southern England, the Jurassic–Cretaceous boundary has long been taken
for convenience at the base of the Hastings Beds although the precise junction
has varied from author to author (see Allen 1955a, p. 266). However, from a
study of the ammonite zones of marine sequences elsewhere and their inferred
equivalents in southern England, Casey (1963, p. 5) has suggested that the
base of the Cinder Bed of the Middle Purbeck marks the base of the Cretaceous
(i.e. Ryazanian) in southern England. He has proposed therefore that the
Cinder Bed and the Purbeck Beds lying above it be called the Durlston Beds

(1963, p. 14) and included in the Wealden (see also p. 27). Casey (1964) has suggested a date of 136 m.y. for the Cinder Bed. Dating of the younger Wealden is less precise. It is generally agreed that the Weald Clay is pre-Aptian (Casey 1961, p. 490) although Hughes (1958) placed part of the Weald Clay in the Aptian (see footnote *in* Casey 1961, p. 490 for re-interpretation of this correlation). The Weald Clay probably spans the Barremian and some (Anderson *in* Kaye 1966) or all (Allen 1955a, b; 1959; 1965) of the Hauterivian. Accordingly the Hastings Beds must be in the main pre-Hauterivian, i.e. Valanginian in age with some of the lower beds of Ryazanian age (Casey 1963, p. 13; Allen 1967a, p. 60). A date of 118 m.y. is tentatively suggested by Casey (1964, p. 199) for the top of the Hauterivian.

Work by Allen (1938–67) during the last 30 years has built up a picture of the growth and subsidence of successive delta complexes in the Weald. He recognized three major cyclothems; the Ashdown Beds and Wadhurst Clay comprise the first; the Lower Tunbridge Wells Sand and Grinstead Clay the second; and the Upper Tunbridge Wells Sand and Weald Clay the third. Where the Cuckfield Stone is present the second cyclothem can be divided into two minor ones, the lower formed by the Lower Tunbridge Wells Sand and Lower Grinstead Clay, and the upper by the Cuckfield Stone and Upper Grinstead Clay. Where the Grinstead Clay is absent in the east it is impossible to draw a line between his second and third cyclothems.

Each major cyclothem can be made up of several distinct horizons, although locally an horizon may be absent or duplicated several times (Allen 1960a). The lower part of each cyclothem is dominantly sand or silt whilst the upper and usually relatively thin part is clay with limestone. Generally the sequence within a major cyclothem in the Weald can be summarized:

graded passage or sharp break with erosion	
Clays and thin limestones, commonly with *Equisetites* soil beds, *Neomiodon* shell-beds and ostracods; thin lenticular sandstones particularly near the base; towards the top dark ostracod-rich clays tend to dominate	Argillaceous (clay/lime-stone) part of cyclothem
Pebble-bed, thin, graded	
sharp break with erosion	
Sandstones dominant and becoming massive near the top; replaced or overlain by argillaceous sandy siltstones in some areas	Arenaceous (sand/silt) part of cyclothem
Siltstones and silty clays with subordinate sandstones	

The phases of the cyclothem are thought to correspond to pulses of the Neocomian sea (Allen 1959, p. 341–2).

The most recent zonal scheme of ostracod faunas for the Hastings Beds is that by Anderson (*in* Anderson and others 1967, p. 189), who recognizes three ostracod zones for the Hastings Beds: Zone of *Cypridea brevirostrata* corresponding to the lower part of the Ashdown Beds; Zone of *C. paulsgrovensis* including the upper part of the Ashdown Beds and the lower part of the Wadhurst Clay; Zone of *C. aculeata* embracing the upper part of the Wadhurst Clay and the whole of the Tunbridge Wells Sand. He has shown that it is possible to subdivide the *C. paulsgrovensis* and *C. aculeata* zones into a number of beds separated by 'marine' bands (1967, p. 193, pl. xiv). As the lithology has not generally favoured the preservation of ostracods the Ashdown Beds and Tunbridge Wells Sand cannot be subdivided in such detail (Anderson *in* Shephard-Thorn and others 1966, pp. 83, 84).   C.R.B., R.A.B., R.W.G., E.R.S.-T.

ASHDOWN BEDS

This lowest member of the Hastings Beds occupies approximately 50 per cent of the map. The largest outcrop, some 70 square miles (180 sq km), is that along the line of the Crowborough Anticline and includes the area from which the formation takes its name, Ashdown Forest.

Calculations from the field evidence and from the Ashdown boreholes (see p. 54) show that the Ashdown Beds are approximately 750 ft (228·6 m) thick in the Crowborough area and thin to about 700 ft (213·4 m) in the Broadoak area. The thickness of 697 ft (212·4 m) recorded in a borehole on the adjacent Tenterden (304) Sheet and previously thought to be due to duplication by reversed faulting (Shephard-Thorn and others 1966, p. 27) is probably the true thickness. Approximately 750 ft (228·6 m) are present in the Penshurst district on the Sevenoaks (287) Sheet to the north, and 730 ft (222·5 m) were proved in a borehole at Worth [289 350], 6 miles (9·6 km) to the west of the present district. A borehole at Placelands, East Grinstead [393 384], proved 681 ft (207·6 m) of Ashdown Beds resting on 9 ft (2·7 m) of dark grey clay. This latter may be a clay within the Ashdown Beds or it may be within the underlying Purbeck Beds.

The base of the Ashdown Beds is indicated by a marked increase in the sand and silt content above a thick, predominantly clay, sequence. This lithological change coincides with a palaeontological change shown at this level by the ostracod faunas (Anderson and Bazley 1971), and there is no so-called Fairlight Clay facies at the base of the Ashdown Beds.

Below the base of the Ashdown Beds the stiff grey and sometimes red Upper Purbeck clays contain only thin sandstone and silt horizons. Above this sand and silt are predominant, but there may be a gradual transition upwards from clay to silts and sands over a thickness of between 10 and 20 ft (3·0–6·1 m). This, in many places, makes it difficult to place a precise limit to the clays, particularly as sandy wash from the strata above tends to mask the junction at the surface. The presence of clay ironstone near the top of the Upper Purbeck clays has in the past been the cause of mining near the junction and these old pits may therefore give a clue to its position. Normally the sands and silts of the basal Ashdown Beds make a slight feature above the clay. Springs frequently flow from this junction, although as there are thin sandstone horizons within the Upper Purbeck clays some are also found at lower levels.

The lower Ashdown Beds silts and sands may in places cut down into the top clays of the Purbeck Beds and cause them to vary in thickness over relatively short distances. The Purbeck/Ashdown junction is only exposed in the southeast corner of the area, along the sides of deep valleys cut by tributaries of the River Dudwell. Landslips are very common and a continuing feature along these valleys.

The basal silts and fine-grained sandstones are generally light grey in colour but weather yellow. Silt is the dominant component in at least the lower 500 ft (152·4 m) of the Ashdown Beds. Many harder sandstone beds occur within these silts and make good features, but they could rarely be traced continuously for more than a short distance. To the dominance of silt in the lower two-thirds of the Ashdown Beds is attributed the badly draining soil from which so much of the country underlain by these strata suffers.

D

Clays, sometimes red in colour, were found at several horizons within the Ashdown Beds, but could not be traced far. Where this apparent lack of continuity is not due to sandy wash from above obscuring the outcrop it is probable that the clays are lenticular due either to localized deposition or to channelling, as would be expected under deltaic conditions of deposition (Allen 1959; Taylor 1963). It is notable that clays in the Ashdown Beds were found to be more common in the south-east of the area than elsewhere. Only two clay beds were found in the Crowborough outcrop.

Allen (1959, p. 290) has noted a coarsening north-westwards across the Weald in the lower Ashdown horizons. It should be noted, however, that one of the sandstone localities quoted, the Kent Water Valley west of Cowden, is now known to be in fact Ardingly Sandstone (see p. 77).

The high proportion of silts in the lower part of the Ashdown Beds may account for many logs made by well drillers showing 'clay' for considerable thicknesses in the lower part of the Ashdown Beds. Although often of clayey consistency when wet these rocks on drying are usually found to be true silts, and the well drillers' logs should be interpreted accordingly.

Occasionally layers were found with infrequent small well-rounded pebbles, usually of quartz, and generally associated with cross bedding; they do not therefore mark major periods of erosion.

The sands are chiefly well-rounded quartz and at most horizons contain plant material. A preserved fauna is rare, only occasional casts, usually of poorly preserved bivalves, being found.

It is in the upper part of the Ashdown Beds that the rhythms of thin silty clays passing up to silts and then sands are best developed. The silty clays, often obscured by downwashed sand and silt, have only locally been distinguished on the map. The features made by the massive sandstone at the top of each rhythm could, however, sometimes be mapped, but seldom could they be traced far. As would be expected with deltaic sediments, conditions appear to have varied considerably even over small distances and correlation by these features is not sufficiently certain to be of value.

Only in the top 100 ft (30·5 m) do the sandstones become at all coarse, but the grains seldom exceed 0·5 mm in size (Allen 1959, p. 300) and are usually less than 0·1 mm (Dines and others 1969, p. 35). It has been demonstrated by Allen (1959, p. 291) that the top Ashdown Beds coarsen towards the south-east, a reversal of the trend seen in the lower Ashdown Beds (but see p. 39). Locally, in the eastern part of the area, low crags, similar to those produced by the Ardingly Sandstone, are formed by the Top Ashdown Sandstone (Plate VIIA).

The Top Ashdown Pebble Bed varies in thickness (0 to 5 in (0–127 mm)) and pebble size (see details). Usually it is at the top of the upper massive sandstone, but sometimes it is a little way (about 4 ft (1·2 m)) above, with ripple-marked sands and silts below. When examined in detail there are sometimes two or three distinct pebble beds and above these there is commonly a passage through thin sands and silts into the silts and clays of the Wadhurst Clay. Very often a thin ripple-marked sandstone was found no more than a foot above the pebble bed; such sandstones are common for several feet above the pebble bed.

The pebbles in the Top Ashdown Pebble Bed are described more fully by Allen (1949a; 1960b; 1961; 1962; 1967a, b). Selected details are given in the present account.

The pebble bed is taken as the stratigraphical base of the Wadhurst Clay; it represents the transgression of the pro-delta Wadhurst Clay across the Ashdown Beds surface. Above the pebble bed there is often a passage through silts (up to 15 ft (4·6 m) at Cowden; see p. 46), into Wadhurst Clay. For descriptive purposes and because of its lithological continuity the Top Ashdown Pebble Bed is described with the Ashdown Beds.

Sphaerosiderite is common, particularly in the lower two-thirds of the Ashdown Beds. In the sandstones honeycomb weathering was found in places (see Plates VIA, VIIB) and this, coupled with nodular structures, is thought to indicate an original carbonate cement which has now been leached away. Tests on surface rocks gave no evidence of any remaining carbonate. Jointing in the Ashdown Beds is not usually very well developed; some examples are noted in the detailed account.

The Ashdown Beds crop out extensively in streams and other cuttings, but only the more important or significant exposures are described in detail.

C.R.B., R.A.B., R.W.G., E.R.S.-T.

WADHURST CLAY

The Wadhurst Clay varies in thickness from about 110 to 235 ft (33·5–71·6 m). Topley (1875) described the formation in some detail and noted the occurrence within predominantly grey clays (mudstones) and siltstones, of thin ironstones, calcareous sandstones (Tilgate Stone), sandstones and shelly limestones. He also mentioned the presence of red and green clays particularly in the upper part of the formation (see also Drew 1861).

Apart from work by the Geological Survey and some valuable contributions from members of the Weald Research Committee of the Geologists' Association, the Wadhurst Clay has received little attention. More recently Allen (1941; 1947; 1949a, b; 1959) has worked on the soil beds, bone beds and pebble beds of the Wadhurst Clay.

In 1960, before the survey of this area was started, three boreholes were drilled in the Wadhurst area, collectively providing an almost complete sequence of the Wadhurst Clay which is now taken as one of the reference sections (see p. 32; Anderson and others 1967).

The junction with the Ashdown Beds, commonly picked out by a feature, is usually sharp and marked by the change from coarse sands of the Ashdown Beds within a few feet to grey clays and silts. In the upper few feet of the Ashdown Beds one or more pebble beds may occur (p. 63). A pebble bed may also mark the actual line of change, but frequently sand, silts and silty clays, up to a maximum of 15 ft (4·6 m) thick, were found above the Top Ashdown Pebble Bed. These transitional beds pass up into a thin soil bed with the roots, rhizomes and stems of *Equisetites lyelli* (Mantell) in a position of growth; this is the Brede *E. lyelli* Soil Bed (Allen 1941 and 1947). *E. lyelli* soil beds were found at at least eight horizons in the Wadhurst Park boreholes and their value as stratigraphic marker horizons in the field is limited (Anderson and others 1967, p. 179). The soil horizon is overlain by thin dark clays which are succeeded by a shell bed composed almost exclusively of valves of *Neomiodon* [*Cyrena*] *medius* (J. de C. Sowerby); this is the Brede *N. medius* Shell Bed (Allen 1938; 1949a, b). The shell beds occasionally form thin limestones or clay ironstones and together with the thicker overlying clay ironstone may form a bench-like feature near the base of the Wadhurst Clay.

Very fine-grained sandstones were noted in the Wadhurst Park boreholes at two horizons. Fine sands were found at five other localities in this area, and are described in the details section, but at the surface these could not be traced for more than short distances. The lower of the two sandstone horizons is not thought to be the equivalent of the Hog Hill Sand (see p. 32).

Above the shell bed comes a sequence mainly of shales and siltstones that make up the body of the formation. Another possible marker horizon is the High Brooms *E. lyelli* Soil Bed (Lock 1953), about 30 ft (9·1 m) from the top of the Wadhurst Clay. Allen (1962, p. 230) has demonstrated that the Balcombe *E. lyelli* Soil Bed (Allen 1947, p. 311) is equivalent to the soil bed at High Brooms.

The junction between the Tunbridge Wells Sand and Wadhurst Clay, where seen in section, is sharp and often marked by a spring line. It is usually masked by a thin downwash of silts and sands (Head) from the Tunbridge Wells Sand, but unless the Head proved to be greater than 4 ft (1·2 m) thick it has not been shown separately on the map. The top few feet of the Wadhurst Clay are commonly red-stained. The reddening is thought by Shephard-Thorn and others (1966, p. 26) to be a primary feature resulting from brief emergence and sub-aerial weathering of the delta sediments. This view is not shared by Anderson and others (1967, p. 176) who believe that the reddening is due to secondary chemical activity of waters moving along the interface with the Tunbridge Wells Sand (see p. 65). The presence of red clay, when augering through Head, is a valuable mapping aid for locating the top of the Wadhurst Clay.

In the eastern area springs were frequently found to issue from the top of the Ashdown Beds, as well as from the basal sands of the Tunbridge Wells Sand. Less commonly, springs arise from layers of siltstone or fine sandstone within the Wadhurst Clay; this may possibly explain the constant level of water in some of the old pits dug in the clay.

The country formed by the Wadhurst Clay is naturally very wet and badly draining. Without artificial drainage boggy ground forms and reeds flourish. The soil is heavy, even where drained, and is mostly used for pasture and the dominantly medium grey mudstones generally weather to form a yellow clay subsoil. Old pits, usually flooded, are a feature of country underlain by the Wadhurst Clay; most were either excavated in the lower 50 ft (15·2 m) of the sequence or at the top, where Lower Tunbridge Wells Sand was worked as well as Wadhurst Clay (see Dines and others 1969) but some pits are found at all horizons. The large old pits were probably dug for a variety of reasons; many occur in the lower part of the Wadhurst Clay where the ironstone is most abundant, but as the 'pay horizon' was so thin considerable overburden would have had to be removed and dumped as spoil in working this. It is probable that most pits were in fact opened for brick and tile manufacture; or for clay to use in 'marling' the fields (a practice that apparently formerly was carried out indiscriminately on soils of all types); or to form waterholes for stock. Any iron encountered was probably put on one side and sold separately (Straker 1931, p. 106). Ironstone was usually mined by bell-pit excavation (= mine-pit) (Topley 1875, p. 334). It is possible that some of the thin limestone seams might have been taken for lime, or used as a flux in the furnaces. Bell-pits can still be seen at many places (see details) and are almost always at an horizon 20 to 30 ft (6·1–9·1 m) above the base of the Wadhurst Clay, although ironstones are common in the lowest 50 ft (15·2 m) of Wadhurst Clay (Anderson and others

1967). Cattell (1970) has shown that the distribution of bloomery sites in the south-eastern part of this district is very closely related to the Ashdown Beds/ Wadhurst Clay junction.

The ironstone beds are of two types, nodular clay ironstone and ferruginous limestone. The latter is usually a shelly limestone probably secondarily enriched by iron and is generally less common than the nodular clay ironstone.

The thin limestone beds are usually shelly, containing most commonly the thin shelled bivalve *Neomiodon* [*Cyrena*]. *Neomiodon* occurs throughout the sequence, but the gastropod *Viviparus* [*Paludina*] is less abundant. In the Wadhurst Park boreholes *Viviparus* was found only at five horizons and may become of value as a stratigraphic marker when more sequences are known from these deposits. Fish remains are found all through the Wadhurst Clay.

Most of the sedimentary aspects of the Wadhurst Clay have been described from the Wadhurst Park boreholes; it is suggested that the sediments were deposited under shallow-water lagoonal conditions in which the water varied from generally brackish to, at times, almost fresh-water and, at other times, possibly marine, as evidenced by variations in the proportion of 'marine' species in the ostracod faunas (Anderson and others 1967).

In the detailed description only the more important localities are given. A feature of most of the Wadhurst Clay stream exposures is the high degree of folding of the strata which is usually superficial, being primarily due to valley bulging.   C.R.B., R.A.B., R.W.G., E.R.S.-T.

TUNBRIDGE WELLS SAND

As stated on p. 33 there are two different successions of the Tunbridge Wells Sand of this district. West of a line drawn from Tunbridge Wells to Uckfield a lower and upper division are recognized, separated by the Grinstead Clay. In the east it has not proved possible to correlate any of the impersistent, variable silty clays with the Grinstead Clay (see also Shephard-Thorn and others 1966, p. 64). Correspondingly, no representative of the Top Lower Tunbridge Wells Pebble Bed has been found. In the west the outcrops of the Grinstead Clay and Ardingly Sandstone are closely related and both occur together. Where the Grinstead Clay is absent in the east, massive sandstones do occur about 60 ft (18·3 m) above the base of the formation, but only rarely do they form good crags and the feature formed by beds at this horizon is variable over short distances. Nowhere within the district is the Tunbridge Wells Sand succession complete; the thickness of the undivided sequence is in excess of 230 ft (70·1 m). Reeves (1948, p. 254), in an artificial classification of the Tunbridge Wells Sand at Saxonbury Hill, near Crowborough, incorrectly took Weald Clay as capping the hill and accordingly deduced that the Tunbridge Wells Sand is here 260 ft (79·2 m) thick (see p. 87). The impersistent silty clay seams recognized here and elsewhere within the district are generally grey in colour, sometimes mottled with red. Their outcrop is commonly masked by downwashed sands but may be traced by augering, by features and by springs issuing from the base of the overlying sands. Springs and seepages are an ubiquitous feature of the Tunbridge Wells Sand/Wadhurst Clay junction.

Details of the lithologies and thicknesses for the subdivisions in the west have already been given (pp. 33–6).

The Lower Tunbridge Wells–Grinstead Clay megacyclothem is regarded by Allen (1959, p. 297) as directly comparable to that of the Ashdown–Wadhurst. The upward change from Wadhurst Clay to Lower Tunbridge Wells Sand is sharp. At East Grinstead the "sandy forests advancing directly across eroded Wadhurst clays" (see Topley 1875, p. 83, fig. 13), cited by Allen (1959, p. 298) as evidence for an eroded upper surface to the Wadhurst Clay, possibly represent not an original feature but the result of cambering (see p. 66).

Minor cyclothems are present throughout the sequence; one intensively investigated by Allen (1959, p. 298) occurs near Pembury Hospital. The pellet breccia which caps a massive sandstone and which is overlain by argillaceous siltstone has been tentatively correlated with the Top Lower Tunbridge Wells Pebble Bed. Roots similar to those of the East Grinstead Soil Bed (Allen 1947, p. 310) penetrate the massive sandstone. Washouts, as seen in the Pembury section, are more frequent in the upper part of the Lower Tunbridge Wells Sand (Allen 1959, p. 299). The Top Lower Tunbridge Wells Pebble Bed, where present, is in most respects analogous to the Top Ashdown Pebble Bed. The size and number of exotic pebbles in this upper pebble bed decrease eastwards and westwards of a line running from the East Grinstead–West Hoathly–Ardingly area, where the Top Lower Tunbridge Wells Pebble Bed is at its maximum, to the Eridge and Speldhurst area in the east and to many of the localities in the Balcombe and Haywards Heath area in the west where the pebble bed is absent. Exposures thereabouts show the Grinstead Clay to rest directly on the Ardingly Sandstone (Allen 1959, p. 303, fig. 12).

As in the Wadhurst Clay the Top Lower Tunbridge Wells Pebble Bed, the basal member of the Grinstead Clay, marks the start of the next major transgression. Above the pebble bed thin clays alternate with thin lenses of sandstone and siltstone and pass up into dark shaly clays containing ostracods, *Unio*, *Viviparus* and ironstone lenticles. The basal passage beds are penetrated by numerous *E. lyelli* rootlets (Allen 1959, pp. 301–7). At East Grinstead an *Equisetites* soil bed, the Hackenden Soil Bed (see p. 92) occurs in the middle of the Grinstead Clay. The Balcombe *E. lyelli* Soil Bed, formerly assigned to the Grinstead Clay, has already been cited under Wadhurst Clay (see p. 42).

The Cuckfield Stone within the Grinstead Clay is frequently capped by a pebble bed (the Cuckfield Pebble Bed) which forms a transgressive base to the pro-deltaic Upper Grinstead Clay. In the south the base of the Cuckfield Stone channels down through the Lower Grinstead Clay to rest on Ardingly Sandstone. On the adjacent Horsham (302) Sheet it is possible to trace the progressive cutting out of the lower clay, but where the two sandstone members are in contact, as in the present district, they are indistinguishable in the absence of clear sections. The Grinstead Clay to the south and east of Danehill represents therefore the Upper Grinstead Clay of other areas, and the sandstone immediately underlying it is the equivalent of the Cuckfield Stone and Ardingly Sandstone combined.

The Upper Tunbridge Wells Sand is not fully represented in this district (see p. 43 for comment on the classification of strata at Saxonbury Hill adopted by J. W. Reeves); only the lowermost 150 ft (45·7 m) are exposed. Locally it has been possible to trace certain thin seams of red silty clay, but generally the Upper Tunbridge Wells Sand is devoid of marker horizons. In the area of the Horsham (302) and Reigate (286) sheets the sandstone which forms the top of the Upper Tunbridge Wells Sand is capped by a pebble bed (see p. 36).

For descriptive purposes the account of undivided Tunbridge Wells Sand has been included with the Lower Tunbridge Wells Sand; for many of the outcrops in the east are thin and include mainly the lateral equivalent of the Lower Tunbridge Wells Sand.   C.R.B., R.A.B., R.W.G., E.R.S.-T.

DETAILS

ASHDOWN BEDS

Upper Stonehurst Farm Inlier [430 410]. This small inlier of Ashdown Beds, less than ½ square mile (1·3 sq km), is bounded on the east by the Lower Stonehurst Farm Fault. The faulted junction of the Ashdown Beds and Wadhurst Clay can be seen in the stream bed at 4370 4105. The beds dip 3°S.W. Exposures are rare and are limited to the stream sections, where thin flaggy sandstones are to be seen.

The shallow valley which starts in Old Furzefield Wood [432 413] and extends south-westwards to 430 419 is fault guided, with downthrow of Wadhurst Clay 50 ft (15·2 m) to the north-west.

Blockfield–Hammerwood–Holtye–Cowden. The Ashdown Beds at Blockfield are bounded to the north by the Blockfield Fault and to the south by the Gotwick Farm Fault. The latter fault dies out eastwards, where Ashdown Beds are overlain conformably by the Wadhurst Clay. A borehole at Dormans Park [397 405], on the north side of the Blockfield Fault, penetrated 542 ft (165·2 m) of Ashdown Beds without reaching the base of the formation. The old pit [4060 3975] 200 yd (183 m) N.E. of The Larches lies close to the Gotwick Farm Fault, which has clearly disturbed the Ashdown Beds. At the southern end of the pit, the flaggy, dark brown, silty fine-grained sand-stones dip 15°S.; in the centre these beds are vertical; on the north they dip 20°N. Around Hammerwood [438 392] the Ashdown Beds outcrop is bounded to the south by the Homestall Fault, and to the north by the Blockfield Fault. Wadhurst Clay crops out to the west.

Fragments of the Top Ashdown Pebble Bed are common on the surface just north of Shovelstrode Farm [4195 3815], while sections some 25 ft (7·6 m) below this horizon, in yellow and yellowish white silts, are exposed in the ditches on either side of Shovel-strode Lane [417 379]. The sandpit opposite Shovelstrode Farm [419 379] has now been almost filled in, but the top few feet of siltstone are still visible. Good sections of up to 6 ft (1·8 m) of yellow and grey silts and siltstone are to be found in the sunken lane [4385 3793] leading to Owlett's Farm.

A sharp anticline associated with the Blockfield Fault and exposed in the old pit [441 397] 300 yd (274 m) N.E. of Hammerwood church has already been noted (p. 9).

Thin flaggy sandstones are to be found in the stream bed [4527 3737] on the east side of the unnamed fault that passes through Roughfield Wood, 700 yd (640 m) S.W. of Beeches Farm.

An old pit [456 399] at the northern end of Holtye Common exposes alternating bands of thinly bedded silty sands and more massive bands; it lies 50 yd (45·7 m) S. of the Blockfield Fault.

In Liveroxhill Wood [4460 4135], ½ mile (0·8 km) W. of Leighton Manor Farm, a small fault trending 060° throws the Wadhurst Clay/Ashdown Beds junction down 25 ft (7·6 m) S.E. About 120 yd (110 m) S.E. of this point, the stream cuts into flaggy sandstones that are only a few feet below the top of the Ashdown Beds [4475 4130].

The outcrop to the north of Cowden shows a general dip of just under 2°S.E. A roadside exposure [4675 4085] north-west of Cowden reveals the Top Ashdown Pebble Bed overlying 3 ft (0·9 m) of clayey silt which in turn rest on finer sand. The 3 ft (0·9 m) of clayey silt are interpreted by Allen (1960a) as alluvial topsets formed during the approach of the Wadhurst transgression. Here the basal 15 ft (4·6 m) of the Wadhurst Clay are silts (Allen 1959, p. 295) and the clay proper was only proved in augering 40 yd (37 m) E. of this pit. This relationship of a silty base to the Wadhurst Clay and a silty top to the Ashdown Beds is even more noticeable in the inlier around Holywych House [482 404], where it is very difficult to locate the precise boundary between the two formations. The sunken lane leading to the now demolished Snailgate Cottages [4885 4047], 500 yd (457 m) E.N.E. of Lower Holywych Farm, exposes sporadically 7 to 8 ft (2·1–2·4 m) of flaggy sandstones on the western side of the small stream.

Tye Farm Inlier [4815 3840]. This inlier is bounded to the south by the South Groombridge Fault, which here has a downthrow of about 350 ft (106·7 m) S.; the Upper Grinstead Clay is faulted against the upper Ashdown Beds.

Thin flaggy sandstone is exposed in the bank of the lane [4825 3840] leading east from Tye Farm.

The Groombridge Horst. Ashdown Beds on the horst were formerly mapped as Tunbridge Wells Sand, and as such formed part of the 'Middle Tunbridge Wells Sand' of Milner (1923a; see Bazley and Bristow 1965). The detailed survey has demonstrated that the Ashdown Beds are here at least 530 ft (161·5 m) thick.   C.R.B.

Two small-diameter (3 in (76·4 mm) reducing to 2 in (51·2 mm)) cored boreholes drilled within the Groombridge Horst at Tangier Farm, Frant [5876 3648 and 5929 3659] proved Ashdown Beds to 383 and 243 ft (116·7 and 74·1 m) respectively (see Appendix I). Both sections show a sequence of irregular cyclothems, ranging from about 10 to 25 ft (3·0–7·6 m) thick, similar to those described for the Ashdown Forest area (p. 51) but in which the siltstone and mudstone parts of the sequences are expanded at the expense of the sandstone parts. The thicknesses of the cyclothems themselves are here much less than in Ashdown Forest and it seems likely that these two boreholes start at a level within the Ashdown Beds that is either equivalent to or below the level of the lowest beds cropping out in the Forest.

Both boreholes start at approximately the same stratigraphical level, the soil bed at 114 ft 8 in (34·8 m) in Borehole No. 1 being at 106 ft 10 in (32·6 m) in Borehole No. 2 and the first red mottled clays appearing at about 236 ft (71·9 m) in both boreholes. Detailed correlations can be made between several of the cyclothems in the two boreholes, but because of imperfect core recovery neither borehole provides a complete sequence on its own. Some of the distinctive lithologies recorded in one borehole only, such as the pellet rock at 210 ft (64·0 m) in Borehole No. 1, the rolled (? slumped) sandstones at 88 ft, 96 ft and 142 ft 6 in (26·8, 29·2, 43·4 m) in Borehole No. 2 and the pellet rock at 233 ft 2 in (70·7 m) in Borehole No. 2, may have been lost in coring in the other borehole or may only be of local extent.

Below about 230 ft (70·1 m) sandstones are rare, and red and grey mottled siltstones and mudstones shot with sphaerosiderite form an alternating sequence reminiscent of parts of the Ashdown Beds seen in the cliff sections at Fairlight Cove, near Hastings.   R.W.G.

A brief outline of the lithology of the Ashdown Beds in one of the boreholes at Groombridge (penetrated to a depth of 338 ft (103·0 m)) is given by Harvey and others (1964, 303/89), although the uppermost strata, as noted by Buchan and others (1940), should include Alluvium and other river deposits possibly to a depth of 23 ft (7·0 m).

A sunken lane [524 363] north of the railway line on the Groombridge–Crowborough road exposes good sections in alternating silts and thick yellowish brown sandstones.

An old flooded pit [5375 3665] alongside the Groombridge–Frant road still reveals 2 ft (0·6 m) of thin brown flaggy sandstone above the water line. A pit given by Milner as "¼ mile [0·4 km] south-east of High Rocks", in his "Middle Tunbridge Wells Sand", is believed to be an old pit [5555 3785] alongside the High Rocks–Frant road; all that can now be seen are 18 in (0·2 m) of thin brown flaggy sandstone.

The sandstone in the railway cutting [5515 3795] ½ mile (0·8 km) W. of High Rocks is much thicker; 2 ft (0·6 m) ribs of massive sandstone protrude through the grassy bank.

Recently bulldozed rides and fire breaks, with ditches on either side, in Broadwater Forest [560 370] reveal yellow and brown silty sandstones and grey, white, red and yellow silts.

The old quarry [5635 3645] ½ mile (0·8 km) S.W. of Spratsbrook Farm is almost completely overgrown but for 2 to 3 ft (0·6–0·9 m) of yellow festoon-bedded sandstone (*sensu* Allen 1962, p. 241, fig. 4; see also Plate IVA). Each festoon or lamina is approximately ½ in (13 mm) thick.

Other localities where small exposures of Ashdown Beds can be found include: Hargate Forest [370 573] with exposures of thin sandstones, silts and silty clay in the ditches bordering the rides; a degraded pit [5873 3715] about ½ mile (0·8 km) S.E. of Rumbers Hill has a few inches of brown flaggy sandstone visible on the western face; thickly bedded sandstone in the railway cutting just south of Windmill Farm [595 376] dips 3°S.E. and in the overflow channel at Benhill Mill [6080 3765] where thick flaggy sandstones are exposed and also in the newly dug stream bed running along the north-western side of the silted-up mill-pond; another old pit [6115 3760] which exposes alternations of thin yellow and grey silts with silty sandstone, and in the quarry [6195 3787] 300 yd (274 m) W. of Sunninglye where thicker sandstone dips 2° at 190°. C.R.B.

Lamberhurst. Top Ashdown Beds strata crop out in a small anticlinal inlier ½ mile (0·8 km) S. of Tong Farm. In the bed and banks of the stream [6752 3885] 350 yd (320 m) E.N.E. of Great Coldharbour Farm up to 4½ ft (1·4 m) of yellow and white sandstone and sand are exposed. The overlying Wadhurst Clay rests irregularly on the top sandstone and appears to cut out the Top Ashdown Pebble Bed in places; the latter where present is represented by lenses of coarse sand up to ¾ in (19 mm) thick, showing small-scale cross bedding.

The top Ashdown Beds are exposed again in an old pit [6818 3830] in Hayden Wood, south-west of Tong Lane; 9 ft (2·7 m) of massively bedded, hard, cream, yellow and brown sandstone, with a ½-in (13-mm) ferruginous gritty band at the top probably representing the top pebble bed, are overlain by 6 in (152 mm) of pale grey basal Wadhurst Clay. G.B., E.R.S.-T.

On the northern side of the North Groombridge Fault between Sandhurst [6485 3800] and Court Lodge, Lamberhurst, the higher Ashdown Beds are exposed in the valleys of several small tributaries of the River Teise.

The Top Ashdown Sandstone forms low rocky outcrops (see Plate IIIB) in the fields and stream banks around Sandhurst. C.R.B.

In Clayhill Wood, up to 15 ft (4·6 m) of massive, pale, medium-grained sandstone at the top of the Ashdown Beds form minor gorges in the stream. Above the topmost sandstone in a stream bank exposure [6526 3778] 350 yd (320 m) at 240° from Clayhill, 4 in (100 mm) of soft coarse sand with fine cross-bedded sandstone lenticles, up to 2 in (50 mm) thick, are overlain by 1 ft (0·3 m) of greyish green clay; above this occurs a band of cross-bedded lenticles of medium buff sand, with quartz and chert pebbles ranging between ⅛ and ¼ in (3 and 6 mm) in diameter, which is taken to be the top pebblebed.

Small waterfalls mark the outcrop of the top sandstones in the stream [6608 3776] 520 yd (475 m) at 210° from Lindridge. At Little Owl House [6614 3733] an old quarry shows up to 8 ft (2·4 m) of massive, medium cross-bedded sandstones, with scattered small quartz pebbles near the top, which here mark the top of the formation. Exposures are poor in the streams draining the country between Cooksbroom Wood and Grantham Hall, but the top sandstones give rise to small waterfalls in several places.

An old quarry [6779 3676], 500 yd (457 m) W.N.W. of St. Mary's Church, Lamberhurst, exposes about 30 ft (9·1 m) of the Top Ashdown Sandstone, as detailed below (the top pebble bed was not seen):

	Ft	in	(m)
Sandstone, buff, rather obscured, with layer of rolled clay ironstone pebbles at base .. .. about	2	0	(0·61)
Sandstone, buff, fine to medium, showing trough cross bedding and with occasional pebbles of quartz and clay ironstone .. .. about	4	6	(1·37)
Grey silt and fine sand, cross bedded .. .. about	0	4	(0·10)
Sandstone, buff, trough cross bedded, with quartz and clay ironstone pebbles .. .. about	2	6	(0·76)
Silt, cross bedded .. .. about	0	2	(0·05)
Sandstone, compact, fine, buff, with ferruginous partings at top and base .. .. about	1	3	(0·38)
Interbedded fine ochreous sandstone and grey silt .. about	1	3	(0·38)
Sandstone, fine, buff, cross bedded .. .. ..8 in to	0	10	(0·25)
Interbedded sandstones and silts, cross bedded ..6 in to	0	8	(0·20)
Sandstone, medium, buff, trough cross bedded, with quartz and clay ironstone pebbles in 'channels' .. .. about	2	0	(0·61)
Sandstone, cross bedded, with silt partings	0	6	(0·15)
Sandstone, fine to medium, white to buff, trough cross bedded; quartz and clay ironstone pebbles occur in a 6-in (150-mm) grit seam at the base and also at the base of the many 'scoops' or channels seen in the face .. up to	3	0	(0·91)
Sandstone, interbedded with silts, cross bedded ..6 in to	0	10	(0·15–0·25)
Sandstone, compact, fine, white, cross bedded with quartz and clay ironstone pebbles in the basal 2 in (50 mm) 6 in to	1	0	(0·15–0·30)
Sand and silt, cross bedded	0	4	(0·10)
Sandstone, fine, white with rough cross bedding showing many intersecting 'scoops' or channels with quartz and clay ironstone pebbles at their bases .. .. up to	3	0	(0·91)
Sandstone, massive, fine, white, cross bedded .. up to	6	0	(1·83)

Other poorer exposures in high Ashdown Beds sandstone occur nearby in the road and lane banks and small old quarries.

The top Ashdown Beds crop out, as a thin rim to the Alluvium of the Teise, north of the South Groombridge Fault near Hoathly; occasionally they form low rocky outcrops. An old roadside quarry at Hoathly [6566 3668] displays the Ashdown Beds –Wadhurst Clay junction: up to 5 ft (1·5 m) of massive fine cross-bedded sandstone at the base are overlain by a 4-in (100-mm) seam of fine sand, with quartz and chert pebbles, ⅛ to ¼ in (3–6 mm) in diameter, taken to be the top pebble bed; this is succeeded by an 11-in (0·3-m) fine buff sandstone, above which poorly exposed shales with cross-bedded lenticles of fine sand and greyish green clays (basal Wadhurst Clay) are seen for 3 ft (0·9 m).

Bewl and Hook River valleys. The top Ashdown Beds were penetrated by several boreholes drilled as part of a recent site investigation for a proposed reservoir approximately 2 miles (3·2 km) N. of Ticehurst. Results proved a valley bulge, which brings Ashdown Beds close beneath the valley floor south of the Chingley Fault (Shephard-

A. Festoon bedding in Ardingly Sandstone at Plumyfeather
Corner, near Withyham

PLATE IV

B. Sandpit in Ashdown Beds, Eridge Park

Thorn and others 1966, pp. 20–1). The record of one of these boreholes sited on the one-inch sheet boundary is given below:

Ticehurst Reservoir No. 4, [6819 3363]. 1961–2. O.D. 156 ft (47·5 m)

	Thickness			Depth		
	Ft	in	(m)	Ft	in	(m)
Wadhurst Clay:						
No core taken	30	0	(9·14)	30	0	(9·14)
Ferruginous bivalve shell bed	0	1	(0·03)	30	1	(9·17)
Grey silty clay with thin cross-bedded lenticles of siltstone bearing rootlet impressions; core much disturbed and broken	1	2	(0·36)	31	3	(9·53)
Ashdown Beds:						
Fine- to medium-grained, white, cross-bedded sandstone with green glauconite grains and small fragments of carbonaceous material spread on cross-bedding laminae which are inclined at about 5° to the horizontal. (No sign of the top pebble bed was seen)	0	5	(0·13)	31	8	(9·65)
Fine glauconitic sand	0	2	(0·05)	31	10	(9·70)
Fine, cross-bedded, glauconitic sandstone ..	0	6	(0·15)	32	4	(9·85)
Core lost	0	8	(0·20)	33	0	(10·06)
Fine, cross-bedded glauconitic sandstone with fine carbonaceous debris	1	9	(0·53)	34	9	(10·59)
Grey, micaceous silty clay, with some fine sand	0	10	(0·25)	35	7	(10·84)
Fine white sand interbedded with grey silt ..	1	0	(0·30)	36	7	(11·15)
Grey silt	1	10	(0·56)	38	5	(11·71)
Fine cross-bedded silty sandstone; core broken	2	0	(0·61)	40	5	(12·32)
Fine, white, cross-bedded sandstone with occasional silty partings; much fine carbonaceous debris and fine colour banding	3	0	(0·91)	43	5	(13·26)
Greyish brown silt with thin siltstones	0	6	(0·15)	43	11	(13·41)
Core lost	2	7	(0·76)	46	6	(14·17)
Grey silt interlaminated with fine sand	0	5	(0·13)	46	11	(14·3)
Compact, grey and white, fine silty sandstone becoming more silty downward and passing into	2	7	(0·79)	49	6	(15·09)
Compact grey silt with thin cross-bedded laminae of fine sand	7	0	(2·13)	56	6	(17·22)
Soft grey silt	0	6	(0·15)	57	0	(17·37)
Fine, white, cross-bedded sandstone with shaly and silty partings	2	10	(0·86)	59	10	(18·24)
Core lost	0	8	(0·20)	60	6	(18·44)
Fine white silty sandstone with greyish green silty laminae passing down into	1	10	(0·56)	62	4	(19·00)
Laminated grey silt and fine sandstone	2	9	(0·84)	65	1	(19·84)
Soft grey silty shale	0	6	(0·15)	65	7	(19·99)
Fine, white, cross-bedded sandstone with shaly laminae; highly variable, some parts predominantly silty	8	3	(2·51)	73	10	(22·50)
Core lost	0	8	(0·20)	74	6	(22·71)
Compact, cross-bedded dark grey silt and fine whitish sand; silt predominates in top 7 ft (2·1 m) becoming more sandy below ..	15	7	(4·75)	90	1	(27·46)
Core lost	0	5	(0·13)	90	6	(27·58)

	Thickness		Depth	
	Ft in	(m)	Ft	in (m)
Fine, white, cross-bedded sandstone with dark laminae of carbonaceous plant debris and silt; rather uniform throughout	9 6	(2·90)	100	0 (30·48)

Shover's Green–Ticehurst. Small faulted inliers of Ashdown Beds occur at Claphatch Bridge [676 317] and near Lower Tolhurst [674 312] north of the Ticehurst Fault, but exposures are poor. South of the Ticehurst Fault and the Withyham Fault the Ashdown Beds crop out in several valleys beneath the Wadhurst Clay which caps the Shover's Green–Ticehurst ridge. In a disused roadside quarry [6616 3115] 350 yd (320 m) at 325° from Holbeanwood up to 8 ft (2·4 m) of massive fine, white cross-bedded sandstones are exposed. The top pebble bed above rests on an erosion surface and is made up, in ascending sequence, as follows: fine white sandstone, with quartz and chert pebbles concentrated in bottom inch and with average diameter about ⅛ in (3 mm), but fair scattering of larger pebbles up to ½ in (13 mm), top 1 to 4 in (25–100 mm) rippled; olive-green silty clay, partly filling ripple hollows on top of bed below, ½ to 1 in (13–25 mm); fine white and ochreous cross-bedded sandstone with quartz and chert pebbles, up to ½ in (13 mm) diameter, in basal 1 in (25 mm), wisps of ochreous clay and a ferruginous rippled top, 2½ to 5 in (64–127 mm). Obscured and weathered basal greyish green Wadhurst Clay overlies the pebble bed. A larger old quarry [6675 3085] 450 yd (411 m) at 80° from Holbeanwood also exposes the Ashdown Beds–Wadhurst Clay junction.

	Ft	in	(m)
Wadhurst Clay:			
Grey silty laminated shales with fine cross-bedded sandstone lenticles .. .. up to	1	6	(0·46)
Pebble bed, coarse sand with quartz and chert pebbles up to 1 in (25 mm) diameter, well cemented: top rippled with olive shale banked against ripples .. .. up to	0	3	(0·08)
Fine, cross-bedded sandstone with occasional pebbles, shaly partings with rootlet impressions .. .. .. 6 in to	0	8	(0·15– 0·20)
Ashdown Beds:			
Fine, white, cross-bedded sandstone with scattered pebbles ..	0	9	(0·23)
Finely laminated sandstone .. .. .. 3 in to	0	3½	(0·09)
Fine white sandstone	1	4	(0·41)
Olive siltstone	0	10	(0·25)
Massive, fine cross-bedded sandstone, laminated; irregular top 'balled up' in manner suggestive of slumping, containing much silt and plant debris in top 1 ft (0·30 m)	3	0	(0·91)
Massive buff and white sandstone, obscured .. .. up to	6	0	(1·83)

E.R.S.-T.

Herontye Inlier. Within the small horst of Ashdown Beds at Herontye two old sandstone quarries still show good sections. The first [4011 3742], 250 yd (229 m) at 040° from Herontye, shows the following section at the junction of the first Ashdown cyclothem (see pp. 31, 51) below Wadhurst Clay with the cyclothem below.

	Ft	in	(m)
Interbedded flaggy and thickly bedded sandstone with thin intercalations of grey silt	5	0	(1·52)
Grey silt	0	6	(0·15)
Medium to coarse sandstone with very coarse lenses and an undulating iron-cemented upper surface	0 to 2 in		(0–0·05)

Ft   in   (m)

Fine- to medium-grained brown and reddish brown massive and
   thickly bedded sandstone containing rare disseminated small
   quartz pebbles up to $\frac{1}{32}$ in (0·8 mm) in size   16   0 (4·88)

The second [3984 3687], 450 yd (411 m) at 198° from Herontye, has worked building
stone from beneath a capping of Wadhurst Clay and shows:

Ft   in   (m)

Wadhurst Clay:
   Grey shales weathering to yellowish brown clay   2   0 (0·61)
Top Ashdown Pebble Bed:
   Ripple lenses of coarse sandstone with a few pebbles, mainly
   quartz, up to $\frac{3}{8}$ in (10 mm) in size ..   ..   .. 0 to $1\frac{1}{2}$(0–0·04)
Top Ashdown Sandstone: ..   ..
   Massive silty greyish brown sandstone   6   0 (1·83)

Other old pits adjacent to the above, in the grounds of Herontye [3998 3707] and
adjacent to the lane [3974 3674] leading to Boyles Farm, worked the same sandstone
bed.

Ashdown Forest. On the western edge of the forest, adjacent to the Wadhurst Clay
outcrop, four cyclothems (numbered 1 to 4 in descending sequence) can be recognized
within the top 200 ft (61·0 m) of the Ashdown Beds. In each cyclothem a thin silt or
silty clay unit is overlain by a thick cross-bedded sandstone unit which is capped by a
pebble bed or by a ripple-marked coarse ferruginous sandstone. Away from the
immediate vicinity of the Ashdown Beds–Wadhurst Clay junction these cyclothems
cannot be mapped with confidence, and it is impossible to relate many of the numerous
small exposures of Ashdown Beds within the Forest to a place in this succession.

An old sandstone quarry [3926 2948], 950 yd (867 m) S. of Westlands, shows the
following section at the base of the 4th cyclothem in which the lowest sandstone is
possibly the top bed of a 5th cyclothem.

Ft   in   (m)

Thickly and flaggy bedded fine-grained sandstone interbedded
   with silty sandstones and siltstones ..   ..   10 0+ (3·04)
Pale grey silty mudstone passing up into siltstone   1   0 (0·30)
Dark grey shale with plant debris   0   6 (0·15)
Coarse ferruginous sandstone ..   ..   0 $0\frac{1}{2}$ (0·01)
Massive medium-grained clean white sandstone   ..   ..   1 0+ (3·04)

A small pit [4006 3215] 220 yd (201 m) N. of Hospital Farm shows clean white fine-
grained sandstone of the 3rd cyclothem whilst on the opposite side of the valley,
north of Horncastle Wood, another old pit [3982 3217] exposes grey silts at the base of
the same cyclothem. The line of outcrop of the silts is marked here, as commonly seen
elsewhere, by a strong spring line. The shallow overgrown workings 200 yd (183 m)
N.E. of Hospital Farm [4016 3209] and 550 yd (503 m) at 120° from Coldharbour
[4022 3252] are in silts and flaggy sandstones at the base of the 2nd cyclothem. They
are probably at the same level as the shallow pits [3980 3304] 350 yd (320 m) N. of
Coldharbour which were presumably worked for ironstone and which show:

Ft   in   (m)

Flaggy orange-brown silty fine-grained sandstone ..   ..   2 0+ (0·61)
Blue-hearted siderite mudstone with dark brown exfoliation
   weathering   1   4 (0·41)
Interbedded orange-brown silt and silty sandstone   1   0 (0·30)
Evenly laminated grey silt with included thin beds $\frac{1}{4}$ to $\frac{1}{2}$ in (6–
   13 mm), of siderite mudstone   4   0 (1·22)
Massive fine-grained greyish white sandstone, patchily iron-
   stained on irregular upper surface; forms floor of quarry ..   1   0 (0·30)

Exposures of thickly bedded and flaggy sandstone in the roadside [3956 3380] 500 yd (457 m) at 340° from Legsheath Farm are within the sandy part of the 3rd cyclothem.   R.W.G.

Fragments of the Top Ashdown Pebble Bed are common on the surface of the fields [4072 3596, 4085 3587 and 4115 3570] south-east of Horseshoe Farm.

The railway cutting west of Brambletye Crossing [416 357] exposes silts and massive Top Ashdown Sandstone.

A ditch section [4308 3583], 700 yd (640 m) N.N.E. of Forest Row railway station, revealed the pebble bed as a coarse-grained sandstone beneath silty Wadhurst Clay. A stream section [4326 3588] 200 yd (183 m) to the north-east exposes a very coarse type of pebble bed, with some of the pebbles exceeding 1 in (25 mm) in length. The pebble bed is overlain by 1 in (25 mm) of fine sandstone which is followed upwards by silts and clayey silts, passing into the normal Wadhurst Clay within about 1 ft (0·3 m). A similar coarse pebble bed is to be found in the adjacent stream to the east [4365 3581].

Thickly bedded brown sandstone was formerly worked in the old quarry [4792 3618] by Hartfield Station. More massive sandstone can be seen at Withyham: by the side of the road [4959 3565]; in the lane [4987 3535] 600 yd (549 m) E.S.E. of the church; at a location [4973 3532] in a sunken lane; and in the overflow channel [4967 3508] of the lake. At the latter locality the beds dip 15° at 070°.

Over Ashdown Forest outcrops are numerous in ditch sections, sunken lanes and most stream beds.

Flaggy sandstone floors the bed of the stream south-east of Bullfinches [535 340 to 540 345] and is also exposed in the old quarry [5335 3460] to the north of the stream. The sunken lane leading northwards from Renby Grange [5335 3325], ¼ mile (0·4 km) W. of Silverlands, to the stream which runs through Hollybridge Wood [5333 3382] is cut into flaggy sandstone which dips 15° at 325° at one point [5338 3375]. A dip of 15°E. was also noted in sandstones in the stream bed [5435 3410] 250 yd (229 m) N. of Copyhold Farm.

Massive Top Ashdown Sandstone is developed around Hamsell Manor and is well exposed in the lane leading up to the Manor [550 341], where it resembles Ardingly Sandstone.   C.R.B.

In Eridge Park, sections of about 11 ft (3·4 m) of cross-bedded Top Ashdown Sandstone are exposed in an old sandpit [5693 3413], 680 yd (622 m) at 069° from Danegate (see Plate IVB). In the same pit was a large sandstone nodule weathering to a spherical shape. It may have been formed by the segregation of carbonate during diagenesis, weathering having later etched out an originally calcitic centre, though all the nodule and surrounding sandstones are now non-calcareous. Possibly some original carbonate was leached out by percolating water.

On the side of the small valley opposite to the above pit is a tunnel through the massive sandstone; it connects with a large circular man-made cavern open at the surface, and exposing about 14 ft (4·3 m) of massive cross-bedded sandstone.

Some 250 yd (229 m) at 020° from Hamsell Manor a thin ½ in (13 mm) pebble bed was exposed with 3 ft (0·9 m) of sandstones in a small stream section [5533 3408]. This is an intra-Ashdown Beds pebble bed, but is not very far below the Top Ashdown Sandstone.

Massive sandstones at the top of the Ashdown Beds, crop out as crags in roads, streams and ditches all over this area. Just north of Stonewall Ghyll, 600 yd (549 m) at 284° from Stonewall, in the side of an old pit [5552 3360] a thin pebble bed at the top of the Ashdown Beds was found.   R.A.B.

The stream south of Wych Cross Place [4130 3175] has numerous exposures in silts and flaggy sandstone. Locally [4087 3121 and 4064 3081] massive sandstone crops out.

A trial borehole by D'Arcy Exploration Company Limited in 1954 at Wych Cross (Harvey and others 1964, 303/151) [4106 3177] situated close to the Crowborough Anticline western crest maximum (Plate II) proved 379 ft (115·5 m) of Ashdown Beds. Some ¾ mile (1·2 km) N. of Pippingford Park a strong feature marks the position of an outlier of the massive sandstone that was exposed in an old quarry [444 319]; the upper part of the section had weathered to a flaggy sandstone.

Massive sandstone was also noted in the stream [4365 3155] that flows east from Wych Cross, although most of the outcrops were of thinly bedded sandstone or silts. Some 450 ft (137·2 m) of Ashdown Beds were penetrated in a well [4320 3015] at Chelwood Vachery (Harvey and others 1964, 303/4). Just over ½ mile (0·8 km) N.E. of this locality the surface of the ground is bare of vegetation [4255 3070] and gully erosion has developed (Plate VB) with incised meanders on the plateau-like surface and, where the slope steepens, a gash 3 to 4 ft (0·9–1·2 m) deep.

Numerous exposures occur in the bed of the Annwood Brook, east of Annwood Farm. Horizontal, ripple-marked sandstone was seen at one point [4256 2768], but some 60 yd (55 m) upstream the dip is 5° at 050°.

At Nutley a seam of lenticular clay up to 25 ft (7·6 m) thick crops out [448 286]. Originally thought to be Wadhurst Clay (see Old Series Sheet 5), the bed was thick enough to be worked in the 1860's for bricks, at the southern end of the inlier [448 284], and another pit [4470 2875] appears to have been later opened alongside the Crowborough road. Probably the presence of ironstone within the clay was the reason for siting the numerous bell-pits on the eastern side of the shallow valley [4482 2856]. The names Minepits Farm (now renamed Marlpitts Farm) [4500 2904] and Sweet Minepits (now renamed Marlpits) [4506 2888] presumably originated at the time of iron working.

An intra-formational pebble bed was found in the old quarry [4525 2811] ½ mile (0·8 km) S. of Marlpits; it was seen to rest on 2 ft (0·6 m) of massive yellowish brown sandstone, and its top, containing pebbles up to ¼ in (6 mm) diameter, was ripple marked. Resting on this surface was 4 ft (1·2 m) of grey silt.

A thick slab of sandstone forms a small waterfall [4617 2667] in the stream that drains south from Spring Garden. Some 140 yd (128 m) upstream thin siltstones and sandstones dip 20°S. A waterfall has also formed over a thick sandstone band in the stream [4638 2974] 600 yd (549 m) E.S.E. of Old Lodge.

Massive Top Ashdown Sandstone, dipping 8° at 200°, crops out in the sunken lane [4777 2706] 400 yd (366 m) W. of Heron's Ghyll. At one point [4775 2702] a cave has been excavated in the sandstone. Wadhurst Clay crops out 200 to 300 yd (183–274 m) S.E. of the lane. Top Ashdown Sandstone, dipping 5°S., was also noted in the stream [486 270] ¼ mile (0·4 km) E. of Heron's Ghyll. A dip of 10° at 170° is recorded from the outcrops of Top Ashdown Sandstone around Shadwell Farm [491 269], 1000 yd (914 m) at 105° from Heron's Ghyll. The sandstone forms a well-defined feature in the fields to the north-west of the farm and Wadhurst Clay is present just south-west of this feature. Top Ashdown Sandstone was again noted in the old quarry [499 271] 750 yd (686 m) at 165° from Chillies. The quarry face has recently been cleaned and now exposes (see Plate VA) in ascending sequence: massive ferruginous sandstone, 5 ft (1·5 m) seen; pebble bed with a ripple-marked upper surface, 4 in (0·1 m) maximum; silts and silty sandstone passing up within about 5 ft into shaly grey Wadhurst Clay. The dip is 4° at 080°. The pebbles in the pebble bed are mostly smaller than $\frac{1}{10}$ in (2·5 mm), but occasional ones were seen to be ½ in (13 mm) in diameter; the length of the ripples measured was 4 in (0·1 m) with an amplitude of 1 in (25 mm).

Lower horizons of Ashdown Beds in this vicinity are finer grained, thinner bedded and frequently show the effects of valley bulging (see p. 111).

In the disused quarry [4697 3205] 150 yd (137 m) N.E. of Gills Lap, the following ascending sequence can still be recognized: massive yellow sandstone with an iron-stained upper surface, 2 ft 6 in (0·8 m) seen; white silt, 3 ft (0·9 m); yellowish brown flaggy siltstone, 1 ft (0·3 m) seen.

There are several old quarries in Ashdown Beds at Jumper's Town, near Hartfield. In one [4695 3295] massive sandstone, 1 ft 8 in (0·5 m), rested on 7 to 8 ft (2·1–2·4 m) of thinly bedded silts and siltstone. A fine-grained pebble bed was present in hollows on the surface of the massive sandstone. Some 7 ft (2·1 m) of yellow and greyish yellow siltstones succeeded the pebble bed.

Alternating sections in massive and flaggy sandstone occur in the stream [498 315] draining north-west from Beacon Hill.

The D'Arcy Exploration Company Limited Ashdown No. 1 Borehole (see p. 128) was sited about ½ mile (0·8 km) E.S.E. of Crowborough Warren [5006 3035]. It commenced at 623 ft (189·9 m) O.D. in Ashdown Beds; the base of the Ashdown Beds was proved at a depth of 479 ft (146 m) (144 ft (43·9 m) O.D.). Unfortunately no core was drawn from the Ashdown Beds but calculations from its proven base, from the adjacent Ashdown No. 2 Borehole and from the surface outcrops indicate that the Ashdown Beds in the area have a total thickness of 750 ft (228·6 m).

A clay seam within the Ashdown Beds was worked for bricks in a pit situated close to St. John's Church, Crowborough Cross (Abbott and Herries 1898, p. 451). This is possibly the same clay seam as that recorded on Crowborough Golf Course (p. 56).

An old quarry [5050 3155], ¾ mile (1·2 km) W.N.W. of Crowborough Cross, has 5 ft (1·5 m) of thick grey flaggy sandstone exposed in the bottom of the pit. Small clay pellets (maximum of 1 by 1¼ in (25 by 32 mm)) having an iron-stained margin are scattered throughout the sandstone. This passes up into 7 ft (2·1 m) of more silty and more thinly bedded flaggy sandstone.

South of the Silverlands Fault [5385 3325] the Lower Tunbridge Wells Sand, Wadhurst Clay and Ashdown Beds dip 5°N. The outcrop is dissected by several streams and there are many exposures of the massive Top Ashdown Sandstone: flooring the stream [5265 3255] ½ mile (0·8 km) W. of Boarshead; in the quarry [5223 3189] 200 yd (183 m) at 300° from Hourne Farm; forming low rocky outcrops [5292 3214] in the wood 600 yd (549 m) N.E. of Hourne Farm; protruding through the turf [5410 3238] and occurring as loose boulders on the wooded banks (Rocky Bank [546 323]) ½ mile (0·8 km) S. of Sandhill Farm. Blocks of coarse Top Ashdown Pebble Bed are scattered on the surface of the field [5355 3210] just west of Limekiln Wood, which is on the dip slope of the Ashdown Beds. C.R.B.

In the east to west lane from Blackdon Hill to Redgate Mill Farm [5530 3246] there are excellent exposures in the Top Ashdown Sandstone. The upper 8 ft (2·4 m) of sandstones seen were ripple marked at several horizons, and below this were massive sandstones that dipped to the south-west down the lane at angles of between 7° and 10°. This dip is probably accentuated by cambering.

At Hornshurst Wood [5550 3150] features of at least four major massive sandstones were traced for short distances, with yellow silts and thin clays between each, and representing in all about 200 ft (61·0 m) of strata. Small outcrops in some of the massive sandstones were seen along the railway cuttings here. The following section was estimated from exposures and augering about 520 yd (475 m) at 310° from Rotherfield Station [5617 3063]:

	Ft	in	(m)
Massive fine-grained sandstone	4	0	(1·2)
Sandstones and silts	8	0	(2·4)
Massive fine-grained sandstone	2	0	(0·6)
Sandstones and silts	10	0	(3·0)
Massive sandstone	4	0	(1·2)

(A10312)

A. Junction of Ashdown Beds and Wadhurst Clay
at Quarry House, High Hurstwood

PLATE V

B. Gully erosion on Ashdown Beds near Wych Cross

(A10301)

In Hoth Wood an old sandpit [5641 3180] revealed the following section with intra-Ashdown pebble beds:

	Ft	in	(m)
Fine-grained massive sandstone	1	6	(0·46)
Cross-bedded sandstone	1	4	(0·40)
Pebble bed	0	0¼	(0·006)
Cross-bedded sandstone	1	3	(0·38)
Ripple-marked sandstone	0	2	(0·05)
Pebble bed	0	1	(0·03)
Medium-grained massive sandstone	2	0	(0·61)

These sandstones are probably about 40 ft (12·2 m) below the top of the Ashdown Beds.

A section in about 20 ft (6·1 m) of flaggy sandstones 220 yd (201 m) at 180° from Highgate Farm near Rotherfield [5562 3033] showed more rippled-marked sandstones, with the main axes of the ripple crests trending N.W. to S.E. This horizon was about 80 ft (24·4 m) from the top of the Ashdown Beds.

Along the junction of the Ashdown Beds and the Wadhurst Clay on the east of this inlier, the Top Ashdown Sandstone was well exposed in many stream and old quarry sections. The Top Ashdown Pebble Bed, resting on 8 ft (2·4 m) of massive sandstone with thin silt partings, was only seen 180 yd (165 m) at 160° from Towser's Lodge [5796 3205].

Old quarries [5660 3085] in the Top Ashdown Sandstone at Town Row Green, 200 yd (183 m) at 100° from Heathfield, near Town Row, revealed 20 ft (6·1 m) of cross-bedded massive sandstones with 4 ft (1·2 m) of flaggy sandstone above.

Near Bletchingley Farm sections along an old road [5798 2996] and down a small stream [5765 2998] showed:

	Ft	in	(m)
Massive fine-grained sandstone with five horizons of ripple marks within the upper 10 ft (3 m)	20	0	(6·09)
Flaggy silty sandstone .. .. about	5	0	(1·52)
Silts and thin clays, light to dark grey .. .. .. about	4	6	(1·37)
Dark grey clay, silty in part .. .. about	6	0	(1·83)

This was one of the few exposed outcrops of the deposits immediately below the Top Ashdown Sandstone. A massive sandstone occurs beneath the clay, but the precise junction was not exposed.

Some 200 yd (183 m) at 050° from Yewtree Farm was an old pit [5718 2979] where the sand is reputed to have been used in the local brick industry. The Top Ashdown Pebble Bed is exposed here:

	Ft	in	(m)
Thinly bedded sandstone	1	6	(0·46)
Pebble bed	0	1	(0·03)
Thinly bedded sandstone becoming more massive near the base	6	0	(1·83)

The old quarry [5672 2952], 370 yd (338 m) at 250° from Yewtree Farm, has exposed the following sequence in the Top Ashdown Sandstone, and indicates the gradual increase in grain size upwards:

	Ft	in	(m)
Massive cross-bedded sandstone with some ripple markings	17	0	(5·18)
Flaggy light grey silty sandstones; bivalve casts noted ..	6	0	(1·83)
Blocky light grey siltstones .. .. about	2	0	(0·61)

In the stream below this quarry there were thin mudstones and siltstones, strongly contorted owing to valley bulging.

E

Some 90 yd (82 m) N. of Rotherfield Hall an old quarry [5430 2910] opened in the upper Ashdown Beds has exposed about 20 ft (6·1 m) of massive cross-bedded sandstones which upwards become more flaggy, probably owing to weathering; 6 ft (1·8 m) from the base is a 3-in (75-mm) clay ironstone. On the side of the hill about 600 yd (549 m) at 074° from Rotherfield Hall is a small sandpit [5482 2914] exposing about 3 ft (0·9 m) of cross-bedded sandstone with a coarse pebble bed about 1 in (25 mm) thick near the top. This is the Top Ashdown Sandstone.

Excellent exposures were found in massive sandstones in the top 100 ft (30 m) of the Ashdown Beds to the east of Rotherfield Hall. Notable are the crags [5327 2909], about 10 ft (3 m) high and showing good examples of honeycomb weathering, 70 yd (64 m) at 160° from Tubwell Farm. Large old sandpits and quarries were found near Sandhill Farm [5302 2854].

In a stream section in Scaland Wood [5228 2780], 540 yd (494 m) at 236° from Owlsbury Farm, the junction of the ripple-marked Top Ashdown Sandstone with the Wadhurst Clay silts and clays above was exposed. The railway cutting immediately north of this was through massive sandstone, about 6 ft (1·8 m) being still exposed [5242 2812].

On the southern limb of the Crowborough Anticline the strata can be seen to be dipping at about 5° at 170°. Many outcrops of fine-grained sand and silt occur in the small streams flowing south of this high ground. Particularly good exposures of flaggy fine-grained sandstones with silts and occasional thin clays can be found along the streams of Slaughter Ghyll [5053 2901], 400 yd (366 m) at 040° from Sweethaws, and The Ghyll [5078 2924], 750 yd (686 m) at 040° from Sweethaws. Cross bedding was common, and several minor scours of sand into finer sediment below were noted. Much lignite was seen within the sands and silts, in places cropping out as coal-like seams up to 3 in (76 mm) thick; but one apparent 3-in (76-mm) seam of lignite in fact consisted of very finely laminated white quartz sand and lignite, in ten distinct layers of white and black.

Near Lord's Well, on the Crowborough Common Golf Course [5100 2965], a thin seam of clay about 220 ft (67 m) from the top of the Ashdown Beds was located by augering and following a line of springs. It is not more than 10 ft (3 m) thick but may persist under a wash of sandy material for a short distance (see p. 54).

At Jarvis Brook [531 297], in the brickworks of the Sussex and Dorking Brick Company Limited, a very good exposure has been made in the Ashdown Beds at an horizon about 400 ft (121·9 m) from their top. Cross-bedded massive sandstones overlie grey and white silts with only thin clay seams. The sequence at the north-west end of the pit [5296 2990] was:

	Ft	in	(m)
Light grey silts with some thin (2 in (51 mm) maximum) clays	8	0	(2·44)
Massive light grey, fine-grained sandstone weathering buff 2 ft to	3	0	(0·61– 0·91)
Alternating light grey silts and fine-grained sandstones ..	8	0	(2·44)
Massive cross-bedded fine-grained sandstone .. .. 1 ft to	8	0	(0·30– 2·44)
Light grey silts with medium grey clay seams not more than 2 in (51 mm) thick	18	0	(5·49)

Measurements are approximate only as there was considerable weathering of the quarry sides, and much talus.

In the upper sandstones sphaerosiderite was very common and in places this iron-carbonate had been weathered out to give a fine honeycomb effect to the sandstone. The brown to yellow coloration of many of the sandstones was due to iron, and in places within the silts small clay-ironstone nodules were noted.

The sandstones near the top of the quarry were in places steeply cambered, dips of 45° N.E. being measured. Plant remains were common throughout the sequence, in places making layers up to 2 in (51 mm) thick. The base of the sandstones frequently showed load casts, and sands slumped with silts forming ball-like structures were found. Cross bedding was very common in the sandstones, as were small scours. A large washout or filled channel was well exposed at the north-westerly extremity of the quarry. Here a massive sandstone, increasing over a distance of 50 yd (46 m) from 1 to 8 ft (0·3–2·4 m) thick, apparently cuts down into the pale silts and thin clays below. Much plant material occurred in the lower 2 or 3 ft (0·61–0·91 m) of this sandstone where the washout was most deeply cut. The section of washout was at least 40 yd (37 m) long (Bazley and Bristow 1965, p. 317).

On the southerly face of the Jarvis Brook Brickworks [5310 2950] similar sequences of massive sandstones and silts were exposed. The structures are generally compatible with a suggested delta slope origin for these deposits (Taylor 1963).

The small inliers around Bestbeech Hill. About 90 yd (82 m) S. of Fright Farm (now called Saxonbury Farm), a stream flows along the upper surface of the Top Ashdown Sandstone, following the dip slope almost exactly. About 6 ft (1·8 m) of ripple-marked massive sandstones exposed here [5867 3210] are typical of the upper part of the top sandstone.

Similar ribbon-like inliers exposing the Top Ashdown Sandstone in the deeply incised valleys of small streams occur at Frankham Wood [5935 3200], 200 yd (183 m) at 005° from Frankham; Frankham Farm [5980 3180], 780 yd (713 m) at 065° from Frankham; Sprayfield Wood [6024 3190], 800 yd (732 m) at 305° from Beggars Bush; and Mab's Wood [6153 3245], about 800 yd (732 m) at 340° from Bestbeech Hill. Along the valley sides of the Frankham Wood and Frankham Farm inliers crags up to 10 ft (3·0 m) high are developed and in several places have been quarried. The inlier at Sprayfield Wood is cut into two by a minor fault, with a throw of about 20 ft (6·1 m), trending N.N.W. to S.S.E.

The railway tunnel ¾ mile (1·2 km) W. of Wadhurst is cut through the Ashdown Beds, and at both ends good sections in massive sandstones are exposed. The general section at the northern end [6253 3249] is:

	Ft	in	(m)
Massive ochreous sandstone with a spring issuing from the base	4	0	(1·22)
Light grey silts and fine-grained sands, the sediment becoming coarser downwards	5	8	(1·73)
Massive fine-grained sandstone	10	0	(3·05)

The southerly cutting [6278 3121] of the Wadhurst tunnel exposes a similar sequence, consisting of 26 ft (7·9 m) of fine-grained sandstones varying from flaggy to massive. These sandstones were formerly assigned to the Wadhurst Clay (Topley 1875, p. 74). The rocks at the northern end of this tunnel dip 2° to 3° N. and those on the south dip at similar angles to the south.

Wadhurst Inlier. The Ashdown Beds on which the village of Wadhurst stands are cut by the main Withyham Fault along their north-east side. Only the upper sandstones are exposed, probably a total of about 80 ft (24·4 m). In an old quarry near Washwell Lane [6381 3106], 750 yd (686 m) at 249° from Stone Cross, the upper beds of rock are exposed:

	Ft	in	(m)
Ripple-marked sandstones with thin silts	3	3	(0·99)
Massive sandstone	3	1	(0·94)
Cross-bedded flaggy sandstone	2	0	(0·61)
Massive cross-bedded sandstone	4	6	(1·37)

Mark Cross to Snape Wood Inlier. The Top Ashdown Sandstone makes an impressive feature all around this inlier, and in several places crags are well developed. Beds on the western side of the inlier are exposed in the sides of the valley cut by the Tide Brook, which rises near Sandyden Wood [5900 3100], 300 yd (274 m) at 150° from Mark Cross. The rocks around Sandyden Wood mostly dip very gently to the south-east. About 70 yd (64 m) at 200° from Renhurst 6-ft (1·8-m) crags of massive sandstone were exposed; the Top Ashdown Pebble Bed lies on the upper surface [5852 3053].

In Coneyburrow Shaw, 140 yd (128 m) at 112° from Earl's Farm, the notable cliff-like scarp is formed by the Top Ashdown Sandstone. An old quarry [5980 3084] in beds at this horizon exposes 30 ft (9·1 m) of massive sandstone, near the surface more flaggy, probably due to weathering. In fallen blocks both the Top Ashdown Pebble Bed and some 10 ft (3·1 m) of ripple-marked sandstone below it were noted.

Four features produced by hard massive sandstones within the uppermost 110 ft (33·5 m) of Ashdown Beds, the topmost marking the Top Ashdown Sandstone, may be traced near Earl's Farm [5968 3085] and Houndsell Place (formerly Earl's Place) [5944 3133] on the hillside above the stream. The intervening beds are silts and silty clays.

Exposed outcrops, particularly of the Top Ashdown Sandstone, are frequent in many small streams and old quarries. Particularly good exposures of the 10- to 20-ft 3·0–6·1-m) crags of the Top Ashdown Sandstone are to be found as follows: in an old quarry 540 yd (494 m) at 043° from Little Trodgers [5925 3042]; along the sides of the stream in Lakestreet Wood [5977 2977], 650 yd (594 m) at 338° from Lakestreet Manor; as a continuation of this section formed by crags in and near the old quarry at Highfields [6020 3008]; in the stream [6071 2950] 900 yd (823 m) at 078° from Lakestreet Manor; and on the valley side [6073 3082] 450 yd (411 m) at 075° from Bassetts. At 350 yd (320 m) at 270° from Tidebrook Manor in an old quarry at the roadside [6120 2952] the following sequence was exposed:

	Ft	in	(m)
Wadhurst Clay:			
Medium grey mudstones and thin silts; fragments of *Equisetites* in upper part	2	6	(0·76)
Pebble bed, variable thickness, on uneven ripple-marked sandstone; iron-pan on upper surface	0	2	(0·05)
Ashdown Beds:			
Massive yellow sandstone	1	0	(0·30)
Light grey silty clay	2	10	(0·86)
Massive yellow sandstone	12	0	(3·66)

The pebble bed is a coarse grit composed mainly of closely packed sub-rounded translucent quartz grains about 1 to 2 mm in diameter.

Along the railway line just south of Scrag Oak to south of Snape Wood, former sections in the Ashdown Beds are now mostly grassed over, but the more massive sandstone beds still stand out, and approximate measurements were taken along the cutting from a place [6405 2943] near Gregory's Pit, about 480 yd (439 m) at 140° from Scrag Oak:

	Ft	in	(m)
Wadhurst Clay:			
Medium grey shales and silts	2	6	(0·76)
Light grey medium-grained sandstone	1	0	(0·30)
Light grey silt	0	2	(0·05)
Pebble bed	0	1	(0·03)
Ashdown Beds:			
Light grey clay and silt	1	0	(0·30)
Sand and silts	9	0	(2·74)

						Ft	in	(m)
Light grey sandstone	..	..	..	..	..	4	0	(1·22)
Sandstone and silts	..	..	..	..	..	10	0	(3·05)
Massive sandstone	..	..	..	..	..	6	0	(1·83)
Silts with occasional thin sandstone layers	..	..	..	20	0	(6·10)
Massive lenticular sandstone	..	..	1 ft 6 in to	4	0	(0·46– 1·22)
Silts and sand	..	..	..	..	..	7	0	(2·13)
Light grey sandstone	..	..	..	..	..	1	0	(0·30)
Silts and sand	..	..	..	..	..	7	0	(2·13)
Flaggy sandstone	..	..	..	..	..	4	0	(1·22)
Silts and sand	..	..	..	..	..	14	0	(4·27)
Massive sandstone	..	..	..	..	..	4	0	(1·22)
Silts and sand	..	..	..	..	..	6	0	(1·83)

The silts and sands were determined by auger. The colour of the weathered rock, light grey to yellow throughout, showed the effects of weathering and oxidation of iron.

In the railway cutting [6332 3019] bordering Snape Wood, about 700 yd (640 m) at 095° from Snape, there was another section:

						Ft	in	(m)
Siltstones	..	..	..	..	..	6	0	(1·83)
Fine-grained massive sandstone	..	..	..	..	1	0	(0·30)
Fine-grained sands and silts	..	..	..	..	4	0	(1·22)
Massive fine-grained sandstone	..	..	..	..	0	8	(0·20)
Flaggy silty sandstone	..	..	..	..	..	6	0	(1·83)
Very hard iron-rich, red weathering sandstone	..	..	1	6	(0·46)
Fine-grained pale grey thinly bedded siltstones..	..	..	20	0	(6·10)

Iron was mined from the Ashdown Beds at Snape Wood from August 1857 until the abandonment of the mine in September 1858 and according to Le Neve Foster *in* Topley (1875) came from three horizons. Two of the iron-rich sandstones that crop out along the railway cutting may have been worked in the past; one was the 18-in (0·5-m) band described in the details above, and another, about 2½ ft (0·8 m) thick, occurred at a slightly lower level.

In Darby's Wood at a location [6468 3040] about 600 yd (549 m) at 070° from Walland, the stream flows down the dip slope (about 8°S.) of a ripple-marked medium-grained sandstone that has clay above and is just at the top of the Ashdown Beds.

R.A.B.

Stonegate Inlier. This inlier is bounded on the north by the major east to west trending Limden Fault. The general dip is about 2° to 4°S. and the total thickness of Ashdown Beds present is about 200 ft (61·0 m). The top 30 to 35 ft (9·1–10·7 m) consist of massive or thickly bedded fine-grained sandstone which commonly contains trough cross bedding, ripple marks, lenses and stringers of small pebbles and fragmentary plant remains. Beneath this, beds of massive, silty fine-grained sandstone are rhythmically interbedded with red, brown and grey mottled silts containing plant debris and thin lignitic bands.

R.A.B., R.W.G.

An old quarry [6800 2681], within the top sandstone, 480 yd (439 m) at 128° from Battenhurst, shows 7 ft 6 in (2·3 m) of massive fine-grained sandstone overlain by 3 ft 6 in (1·1 m) of thinly bedded weathered sandstone. An adjacent quarry [6806 2672] shows a similar section.

Ripple-marked sandstones close to the junction with the Wadhurst Clay are exposed [6741 2751 to 6700 2678] in the bed of the stream 500 yd (457 m) N.W. of Battenhurst.

Other sections in the top sandstone occur in old quarries 300 yd (274 m) E. of Cottenden [6787 2831] where 20 ft (6·1 m) of massive sandstone are present and at 600 yd (549 m) at 072° from Battenhurst [6808 2728] where a few feet of stiff yellowish brown clay overlie 8 ft (2·4 m) of massive sandstone with plant debris and cross bedding with ripple marks picked out by iron staining.   R.W.G.

The southern boundary with the Wadhurst Clay is normal, the pebble bed between the Ashdown Beds and Wadhurst Clay being particularly well developed in two deeply cut streams at Coalpit Wood [6526 2847], about 430 yd (393 m) at 187° from Maplesden, and in Hoadley Wood [6584 2802]. At the latter locality the pebble bed, just over 2 in (51 mm) thick and unusually coarse, is at the base of a 9-in (0·2-m) sandstone and was seen to be graded, the coarse pebbles being at the base. Above the sandstone were 3 ft (0·9 m) of cross-bedded and ripple-marked silts and fine-grained sandstones, and below the pebble bed were 4 ft (1·2 m) of pale grey and yellow silts with thin sandy partings.   R.A.B.

Fairwarp Inlier. There are few exposures of Ashdown Beds in the lenticular shaped inlier to the south of Fairwarp [465 260]. Anticlinally folded massive sandstone, the limbs dipping 5° at 10° and 190° respectively, some 25 ft (7·6 m) below the base of the Wadhurst Clay, crops out in the bed of the stream [4733 2638] ½ mile (0·8 km) at 080° from Cophall Farm.

Hendall Inlier. The fault-bounded inlier around Hendall [475 258] is also poorly exposed. Thin flaggy sandstone dips 2° at 200° in the lane [467 254] ½ mile (0·8 km) at 170° from Cophall Farm. The north-eastern boundary is formed by the Greenhurst Fault. West of Stonehouse, where the valley coincides with the fault, thin flaggy siltstones and sandstones crop out in the stream bed on the south-west (upthrow) side of the fault.

High Hurstwood–Burnt Oak Inlier. This inlier is fault bounded over most of its length by the Sleeches Fault to the south and Burnt Oak Fault to the north. Massive Top Ashdown Sandstone floors the stream [4826 2645 and 4821 2622] ¼ mile (0·4 km) E. of Claygate Farm. The southernmost of these two outcrops dips 10°S., and some 40 yd (37 m) S. grey shaly Wadhurst Clay crops out on the stream bed. Around High Hurstwood the Top Ashdown Sandstone was noted at the roadside [4944 2596]; in the old quarry [4965 2584]; and in the stream [4928 2578 and 4935 2608] to the west of the Crowborough road. The sandstone dips 4° at 190° at the latter point.

A well [4969 2579] 300 yd (274 m) N.E. of the Maypole Inn passed through 10 ft (3·0 m) of Wadhurst Clay into Ashdown Beds and continued in this formation to a depth of 115 ft (35·1 m) (Harvey and others 1964, 303/66).   C.R.B.

In an old quarry [5335 2675] about 200 yd (183 m) at 285° from Pinehurst approximately 9 ft (2·7 m) of flat-bedded massive sandstones are exposed. This quarry was excavated in the side of a very good sandstone feature that can readily be traced around this hillside and is probably the Top Ashdown Sandstone.

About 200 yd (183 m) S. of Stone Mill, crags of massive sandstone up to 20 ft (6·1 m) high, are very well exposed [5440 2635]. These crags, in the Top Ashdown Sandstone, can be followed in a westerly direction along the south side of this valley for almost 1 mile (1·6 km), through Stonehurst Wood and Walsted Wood to Under Rockes [5560 2635] about 700 yd (640 m) at 257° from Horleigh Green Farm. The latter locality, partly due to past quarrying, has magnificent crags about 30 ft (9·1 m) high in the massive Top Ashdown Sandstone (see Plate VIIA). Cross bedding can be seen and associated with this are thin bands of small pebbles which are well rounded and dominantly of quartz, as is the remainder of the sandstone. Honeycomb weathering is another feature of this exposure. In the stream nearby are exposed silts and thin sandstones that occur below the massive beds. These have dips which are very variable and probably principally due to valley bulging.

About 350 yd (320 m) at 118° from Streel Farm red and cream-coloured mottled clays traced by augering [5578 2680] indicate a thin clay seam about 40 ft (12·2 m) below the top of the Ashdown Beds; the mottled coloration may, however, be due to some secondary weathering process connected with the nearby faulting.

Hastingford Inlier. All around this inlier the Top Ashdown Sandstone makes an excellent feature and in several places crags are developed, particularly along the southern margins. In a road section [5230 2599] 200 yd (183 m) N. of Hastingford there are 10-ft (3·0-m) high crags with a thin pebble bed near the top.

In and alongside the valley, about 300 yd (274 m) N. of Broadreed Farm crags of massive sandstone occur, and similar crags are seen [5313 2562], about 500 yd (457 m) at 195° from Huggett's Furnace, where two well developed sets of vertical joints trend 069° to 249° and 145° to 325°.

Downford Inlier. About 100 ft (30·5 m) of strata at the top of the Ashdown Beds are exposed in this inlier. All the streams show many good exposures and are deeply cut. In the stream beds the dips of the rocks were very variable, from nearly vertical to horizontal; though determined largely by valley bulging, proximity to the major east to west trending North Mayfield Fault may have controlled some disturbances. The junction of the massive Top Ashdown Sandstone with the Wadhurst Clay was seen at two localities: in the River Rother at Lumps Wood [5598 2802], 700 yd (640 m) at 301° from Salters Green Farm, and in the stream at Angle Wood [5702 2774], 550 yd (503 m) at 080° from Salters Green Farm. At the latter locality a 1-in (25-mm) pebble bed was found at the junction.

In the Ashdown Beds below their uppermost 20 ft (6·1 m), thinly bedded sandstones and a higher proportion of silt are exposed, acutely folded, in the bed and banks of the River Rother; just north of Lymley Wood [5677 2743] is a perfect anticline with its axis trending parallel to the valley sides. Similar acute structures were exposed in the stream just east of Downford [5767 2717] and can be attributed to valley bulging.

Coggins Mill. Approximately the upper 150 ft (45·7 m) of typical Ashdown Beds sands and silts, dipping at 1° to 2° S.E., are found in this inlier, exposed particularly in the many deep-cut streams.

At the junction with the Wadhurst Clay a pebble bed 1 to 2 in (25–50 mm) thick, was found at several localities. These included Six Acre Wood [6176 2757]; near Watlings Wood [5990 2884]; and south of Batts Wood [6269 2720 and 6389 2692].

The Top Ashdown Sandstone has been quarried at several places, i.e. near Great Wallis Farm [5828 2836]; in Clay's Wood [5834 2804]; north of Brick Kiln Wood [5856 2985]; and near Newbridge Wood [6389 2689]. In Heronry Wood [5911 2867], about 300 yd (274 m) at 205° from Mayfield College, 20 ft (6·1 m) of fine-grained cross-bedded sandstone with occasional layers of poorly preserved plant remains are exposed in an old quarry. Near the base is a thin pebble bed.

Below the upper 30 ft (9·1 m) of Ashdown Beds in this inlier silts become more common, but in no place was more than a few inches of clay found. In the main stream at Coggins Mill [5970 2790], and for at least 800 yd (731 m) upstream, exposures of the lower Ashdown Beds were common; thin sandstone and silt predominate. Most sections showed contorted strata, and minor anticlines with their axes trending parallel to the valley sides indicate probable valley bulging.

The following descending sequence was recorded in the stream [5945 2799] 450 yd (411 m) at 320° from Bassett's Farm: massive yellow sandstone 6 ft (1·8 m); flaggy sandstone and silts, 2 ft 1 in (0·6 m); reddish grey clay ironstone, 1 ft (0·3 m); massive yellow sandstone, 3 ft (0·9 m). This was one of the few sections in which ironstone was found in strata of the upper Ashdown Beds.

Shepherd's Hill–Burwash Common Inlier. This inlier is bounded along the north by
the east to west Burwash Common Fault. The southerly part of the inlier is on the
Lewes (319) Sheet. The total thickness of the Ashdown Beds in this area is about 700
ft (213·4 m), but the lack of marker horizons within this division and the complex
zone of faulting along the northern margin of the inlier make precise measurement
difficult. A saddle in a ridge [5590 2210] 600 yd (549 m) at 155° from Dudsland Farm
can be linked by an almost east to west line to two other similar saddles, at Rabbit
Burrow Down [5680 2220] 1000 yd (914 m) at 165° from Isenhurst, and at Markly
Wood [5800 2240] 650 yd (594 m) at 235° from Orchard House. These saddles may be
due to differential erosion along a fault plane, but owing to lack of marker horizons no
movement could be proved on either side.

In the west of this inlier the upper part of the Ashdown Beds is exposed. Crags
of the massive Top Ashdown Sandstone are up to 20 ft (6·1 m) high about 650 yd
(594 m) at 060° from Shepherd's Hill, near the ancient Moat [5206 2251]. Most of the
inlier consists of silts and fine sands, the harder sandstone beds forming features with
spring lines near their base.

At Dudsland Farm [5664 2259] a small temporary section revealed about 6 ft
(1·8 m) of light grey silts with thin clay seams not more than 2 in (50 mm) thick.
Similar light grey silt was augered along the base of this valley, and in the valley to
the north, and is closely comparable to the silts in the middle of the Ashdown Beds
found at Jarvis Brook Brickpit, Crowborough. The silt and sand nature of the middle
Ashdown Beds is shown by the soils and small outcrops in the deep-cut streams in
Tilsmore Wood [5740 2220] and Markly Wood [5810 2270] about 2 miles (3·2 km) W.
of Broadoak.

In the railway cutting [5847 2415] 500 yd (457 m) at 219° from Old Mill Farm about
7 ft (2·1 m) of massive sandstone are exposed. Its top is ripple-marked, possibly
indicating that it is in the upper part of the Ashdown Beds sequence. The ripple marks
here were found to measure 1 in (25 mm) in height and have a wave length of 8 in
(0·2 m), the axes of the ripples trending north-east to south-west.

The clay band in the Ashdown Beds about 1 mile (1·6 m) W. of Broadoak could not
be separately traced farther than shown, although lines of springs [5992 2225, 6035
2262 and 6055 2272] to the east of this locality are indicative of a clay, or silty clay,
at this horizon. A landslip of Ashdown Beds at Broadview Estate involved a thin
blue silty clay (see p. 110). In a trench section near Tanyard Cottage [5895 2200], 920 yd
(841 m) at 189° from Bodell's Farm, a massive fine-grained light grey sandstone,
about 8 in (0·2 m) thick, was exposed beneath medium grey clay, which was not more
than 10 ft (3·0 m) thick and graded up into light grey silts and massive sandstones
which cap the ridge. This thin clay seam is estimated to be about 290 ft (88·4 m)
above the base of the Ashdown Beds. In the stream that rises near Tanyard Cottage
massive sandstones are exposed, and about 120 yd (110 m) at 303° from Bodell's
Farm [5897 2284] about 20 ft (6·1 m) of vertically folded light brown massive sandstones and yellow silts mark either a line of valley bulging or a minor structure
associated with the major fault about 250 yd (229 m) N.

In the stream referred to above, north of Button's Farm Fault there are medium
grey silts and clays dipping at 2° to 3°N. A good section [5982 2365], about 950 yd
(869 m) at 008° from Bodell's Farm, showed about 10 ft (3·0 m) of dark grey silts
with clays overlain by light coloured silts and thin sandstones. These clayey horizons
were only identified in the bottom of the valley in the stream sections. They occur
about 50 ft (15·2 m) below the top of the Ashdown Beds and are probably equivalent
to the clays at similar horizons noted near Bletchingley Farm, Rotherfield, about 4
miles (6·4 km) away.

From Broadoak to the east of this inlier the strata generally dip in a northerly
direction from 2° to 8°. The junction of the Purbeck Beds with the Ashdown Beds
occurs in this area but is never fully exposed.

The fact that Upper Purbeck clays vary in sand and silt content and grade upwards into silts and sands makes a precise placing of the field boundary with the Ashdown Beds difficult. This was taken at the highest true clay horizon; but in Tottingworth Park [6170 2217], for example, two clay beds each up to 10 ft (3·0 m) thick were traced above the Upper Purbeck clays and a clay pit in one of these [6210 2234] may possibly have been dug for ironstone.

In the area between the inlier of Purbeck Beds and the Burwash Common Fault the Ashdown Beds consist dominantly of silts and fine-grained sands. Some yellow, or locally red, clays were found at several places: near Burraland [6240 2323] clay caps the hill and is about 320 ft (97·5 m) above the base of the Ashdown Beds; thin clays traced near South Binns [6253 2400] and Burwash Common [6400 2350] are at similar levels about 150 ft (45·7 m) beneath the base of the Wadhurst Clay; the clay around Street End [6060 2362], about ¾ mile (1·2 km) N. of Broadoak, is possibly only about 70 ft (21·3 m) below the base of the Wadhurst Clay.

In a stream section [6017 2343] 520 yd (475 m) at 303° from Street End nodules of clay ironstone up to 4 in (0·1 m) thick were seen in about 3 ft (0·9 m) of light grey silts. Many small sections of plant-rich sandstones and silts are exposed along this stream.

In Marlpit Shaw [6505 2351] 330 yd (302 m) at 354° from the inn at Burwash Weald, the following sequence was recorded:

	Ft	in	(mm)
Wadhurst Clay:			
Yellow to grey shales and clay with ironstone nodules up to 1 in (25 mm) thick and thin *Neomiodon* limestone	1	6	(457)
Fine-grained ripple-marked sandstone	0	0½	(13)
Pebble bed .. .. ¼ in to	0	1	(6–25)
Ashdown Beds:			
Light grey silty sand	0	6	(152)
Hard yellow sandstone	0	4	(101)

Bateman's Inlier and small inliers north of the Burwash Common Fault. Small inliers occur along streams near Bigknowle [6256 2474], Climshurst Wood [6360 2455] and Furnace Gill [6499 2407]. At the Bigknowle Inlier the Top Ashdown Pebble Bed, up to 2 in (51 mm) thick, was found.

In Furnace Gill, 750 yd (686 m) at 260° from Woodlands, the following section [6499 2407] was obtained:

	Ft	in	(mm)
Wadhurst Clay:			
Grey shales, becoming silty near base	1	8	(508)
Grey fine-grained flaggy sandstones	1	10	(559)
Grey sandstone; washout in places .. .. .. ¼ in to	0	9	(6–229)
Grey shale	2	9	(838)
Pebble bed .. .. 0 to	0	0½	(0–13)
Light grey silt	0	1	(25)
Dark grey carbonaceous silt	0	0½	(13)
Pebble bed .. .. ¼ in to	0	1	(6–25)
Grey clay	0	0¼	(6)
Grey ripple-marked sandstone	0	0¼	(6)
Pebble bed .. .. 2 in to	0	5	(51–127)
Ashdown Beds:			
Grey sandstone, weathers brown	0	4	(101)
Grey shale	1	9	(533)
Light grey silts	2	0	(610)

This section shows pebble beds at three horizons separated by thin layers of silt and clay. One of the silt partings was full of finely broken carbonaceous fragments and appeared black in the section.

Around Bateman's [6710 2380] the boundary between Ashdown Beds and Wadhurst Clay could be easily traced by the well-developed Top Ashdown Sandstone feature. This sandstone has been quarried at several places and Bateman's itself was mainly built from local sandstone, some of which probably came from the small quarry nearby [6719 2390]. The many paving stones used in the gardens of Bateman's are of a Purbeck Beds limestone, probably from the Greys Limestones which were once mined locally.   R.A.B.

South of Dudwell [674 233] good sections in the lower beds are rare, but numerous small exposures occur in the forestry roads.   R.W.G.

WADHURST CLAY

Upper Stonehurst Farm [430 410]–**Cowden–Blackham.** Wadhurst Clay surrounds the Stonehurst Farm Inlier of Ashdown Beds. East of the Beeches Fault, and north of the Kent Water Valley, the Wadhurst Clay has a large outcrop studded with numerous flooded pits; it is about 160 ft (48·8 m) thick.

Thinly bedded grey shaly clay with an ironstone band (close to the base of the Wadhurst Clay) is exposed in the stream bed [4341 4061] at Lower Stonehurst Farm. A series of bell-pits, some 25 to 30 ft (7·6–9·1 m) above the base of the formation, can be traced through the woods (Minepit Wood [4455 4145] and Liveroxhill Wood [4490 4115]) west of Leighton Manor. A small fault located between these two woods throws the base of the Wadhurst Clay 25 ft (7·6 m) down to the south-south-east.

This area was formerly a very important centre of the Wealden iron industry. Bloomeries are recorded at Cinder Wood [435 397], ¼ mile (0·4 km) N.E. of Hammerwood church; Basing Farm [4375 4023]; Beeches [435 413]; Birchenwood and Waystrode [about 458 406] (see description below). A furnace site is at Cowden (Furnace Mill [4550 3992]) and a forge and furnace at Scarletts (Straker 1931; Sweeting 1944). The hammerpond at Scarletts has now silted up (see Sweeting 1944, pls. 2, 3 and 4) but much of the Furnace Pond at Furnace Mill remains open water. It has been suggested that some of the "bloomeries" listed by Straker (? Beechenwood, Birchenwood, Waystrode, Castle Hill, Brockshill, Blackhill and Cotchford) may have in fact been the cinder used by the Romans in the construction of the London–Lewes Way (Straker and Margary 1938, p. 58).

Small faulted inliers of Wadhurst Clay crop out low down in the Kent Water Valley [4370 4018, 437 401 and 4460 3993]. At the first of these localities brick-red clay was augered [4364 4016] just beneath the Lower Tunbridge Wells Sand. Red clay at the same horizon was also augered in the field [4212 4079] 500 yd (457 m) S.W. of Upper Stonehurst Farm.

The Wadhurst Clay to the north of Cowden and around Saxbys dips 2°S.E. Its surface is extensively pitted by old working but there are now no exposures within the clay. The road cutting [4680 4085] ¼ mile (0·4 km) N.N.E. of Cowden church exposes the Top Ashdown Pebble Bed (see p. 46) beneath the 15 ft (4·6 m) of silts and siltstones which intervene between the pebble bed and the stiff clay of the Wadhurst.

Around Blackham the Wadhurst Clay appears to be about 130 ft (39·6 m) thick although landslipping, and possibly cambering (see p. 109), have affected the outcrop: the Lodge Field Farm outlier of Lower Tunbridge Wells Sand dips 4°E.; the Highfields Park outlier dips south-eastwards in the western part and in a general northerly direction in the east; and around Lower Holywych Farm and Hethe Place [480 402] where the junction with the Ashdown Beds is gradational (see p. 46) the dip is about 1½°N.E.

'*Cyrena*' limestone is present towards the base of the clay and fragments are common on the surface of the fields [4767 3983 and 4791 4024] 250 yd (229 m) N.W. and 600 yd (549 m) N. of Hethe Place. '*Cyrena*' limestone at a higher horizon was found in the field [4960 3892 and 4966 3900] 600 yd (549 m) at 210° from Pound Farm. Possibly limestone was dug in the small pits at the south-western corner of the field. Minepit Wood [492 381], ¼ mile (0·4 km) S.E. of Beechgreen House, is indicative of former iron-working. There is much iron slag in the bed of the small stream [5075 3913] 400 yd (366 m) N. of Ashurst Station. This lies close to the site of an old furnace (Straker 1931, pp. 231–2). Straker also recorded a forge sited on the Kent Water ⅞ mile (1·4 km) N. of Ashurst church.

Wadhurst Clay outcrops in the river cliff [5070 3961] 350 yd (320 m) S. of Willett's Farm and poor exposures can also be found in the old gravel pit just north of this locality [5065 3972].

On the eastern side of the Medway there is a narrow outcrop of Wadhurst Clay between the Lower Tunbridge Wells Sand and the Alluvium. Over much of this area the clay is obscured by downwashed sand or terraced river gravels.

Undercutting of the river cliffs has caused small-scale slipping [5160 4005] 500 yd (457 m) at 245° from Chafford Farm. Red clay at the top of the Wadhurst Clay was proved by augering in the field [5147 3932] 200 yd (183 m) S.E. of Chafford Park, and also in the wood [5380 3806] ¼ mile (0·4 km) at 120° from Top Hill.

'Beech Green Stone', formerly dug 2 miles (3·2 km) N.N.E. of Hartfield, is not an horizon within the Wadhurst Clay as Topley (1875, pp. 89, 379) thought but is the Cuckfield Stone within the Grinstead Clay (see p. 92). C.R.B.

East Grinstead–Ashurstwood–Hartfield. Boreholes at Dormans Park, [397 405] and at Placelands, East Grinstead [393 384], proved the total thickness of the Wadhurst Clay to be 240 and 235 ft (73·2, 71·6 m) respectively. Although both records are based on drillers' logs they agree well with structure contours which indicate that the Wadhurst Clay is about 220 ft (67·1 m) thick in this area. South-eastwards, near Ashurstwood, the total thickness is probably about 160 ft (48·8 m).

A clay-ironstone seam is well developed near the base of the formation and has been worked in numerous open pits and bell-pits in the area. Numerous deep pits near the top of the formation may indicate the presence of another clay ironstone at that level. Beds of '*Cyrena*' limestone are common in this area and loose fragments of rhizomes of the horsetail *Equisetites lyelli* (Mantell), probably derived from the High Brooms Soil Bed, have been found at several localities.

At the base of the formation the Top Ashdown Pebble Bed is well developed in this area. Where not obscured by downwash the highest part of the formation was everywhere seen to be reddened.

A narrow strip of Wadhurst Clay crops out between the Gotwick Farm Fault and the Lower Tunbridge Wells Sand outcrop. In Swite's Wood deep pits [3968 3982 and 4019 3981] were dug for marl or ironstone, possibly both, close to the junction with the Lower Tunbridge Wells Sand. Pits at a similar level occur near East Court [4002 3889 and 4006 3846] and south of East Grinstead church [3972 3767 and 3996 3779], the two former localities exposing red clays. Similar red clays occur in the railway cutting [4008 3790] 500 yd (457 m) at 110° from the church, which shows: Lower Tunbridge Wells Sand comprising interbedded flaggy silty sandstones and mottled (catsbrain) orange and grey silts, 6 ft (1·8 m); gap, the junction obscured by downwash, 5 ft (1·5 m); Wadhurst Clay showing small slips in red clay with included shale debris, 2 ft (0·6 m). It is interesting to note that Topley (1875, p. 83), who described this section when it was but little weathered, records "at the junction [of the Lower Tunbridge Wells Sand and the Wadhurst Clay] there is no red clay, but green and blue shale and clay". Thick red clays are very rarely absent from the surface outcrops of the topmost Wadhurst Clay, but in the present-day working pit at High Brooms, Southborough

[594 418], on the adjacent Sevenoaks (287) Sheet, the junction with the Lower Tunbridge Wells Sand is marked by up to 6 in (152 mm) of red clay, although locally this is absent and the sands rest directly on bluish grey shales and mudstones (Dines and others 1969, p. 41). It is probable that the reddening is due, at least in part, to weathering processes which oxidize the ferrous iron of the greenish grey shales to ferric iron in the red clays (see p. 42).

In the same cutting [4002 3802], north of the Eastbourne road, Topley recorded (1875, p. 83, fig. 13) "a curious instance of thinning away of beds". This section is now grassed over but from his figure it seems possible that this feature might be due to cambering (see p. 44) rather than to current action.

A maximum of about 20 ft (6·1 m) of Wadhurst Clay caps the Ashdown Beds in the Herontye Inlier, where the junction of the two formations is marked by the Top Ashdown Pebble Bed (see p. 50). R.W.G.

Several sections in valley-bulged shales occur in the stream course 450 yd (411 m) S.E. of Herontye. The best section [4011 3697] shows: dark bluish grey shales with ostracods and bivalves, 2 ft (0·6 m); calcareous 'blue-hearted' siltstone, 1½ ft (0·5 m); dark bluish grey shales as above, 1 ft (0·3 m); 'Cyrena' limestone band, 1 in (25 mm); dark bluish grey shales as above 1 ft (0·3 m).

Siltstone casts of rhizomes of E. lyelli occur loose in the stream bed here and may have been derived from a soil bed at the level of the High Brooms Soil Bed. Fragments of E. lyelli were found in the road cutting [419 363] 400 yd (366 m) N.W. of Bramblehurst. Allen (1962, p. 229) noted numerous siltstone lenticles about 30 ft (9·1 m) from the top of the Wadhurst Clay associated with rootlets, and rhizomes and ascending stems of E. lyelli. He equated this horizon with the High Brooms Soil Bed. An unusually thick bed of 'Cyrena' limestone, up to 5 in (0·1 m) thick, was found as soil debris [4012 3600] 550 yd (503 m) at 052° from Busses Farm, where it caps a prominent feature. This bed could be followed for some 200 yd (183 m) before being obscured by soil cover. It is estimated to be about 45 to 50 ft (13·7–15·2 m) above the base of the Wadhurst Clay. The layers of grit and shelly limestone recorded by Topley (1875, p. 82) in the railway cutting near Horseshoe Farm [4053 3605] may be at the same horizon.

The basal ironstone, here some 10 to 25 ft (3·0–7·6 m) above the base of the formation, has been extensively worked by means of bell-pits and shallow open pits in Botley Wood [403 357], Minepits Wood [402 362], Horseshoe Rough Wood [405 364], Minepit Wood [410 360] and High Wood [427 360]. To the east of the Eastbourne road bell-pits occur at the same horizon in Minepits Wood [434 352], ¼ mile (0·4 km) N.W. of Pixton Hill. Some of these workings may have provided ore for the Roman bloomery at Whalesbeech [393 345] and the mediaeval forge at Brambletye [417 353] (Straker 1931, pp. 239–41). R.W.G., C.R.B.

To the east of East Grinstead there is a large relatively flat-lying, partially fault-bounded, outcrop of Wadhurst Clay. There is evidence of widespread working of iron from the Wadhurst Clay in this vicinity: bell-pits were noted in the woods [425 386 and 424 395] to the north and south of Shovelstrode Manor, some 20 to 25 ft (6·1–7·6 m) above the base of the Wadhurst Clay; pits in Minepit Wood [418 392], ½ mile (0·8 km) at 200° from Shovelstrode Manor, appear to have worked a higher seam of iron; a forge is recorded at Bower [about 440 384] by Straker (1931, p. 229); much iron slag is to be found in the stream [3927 4330] 500 yd (457 m) at 290° from Bower Farm; and the name Hammerwood [438 392, 4420 3885 and 447 387] is indicative of the former industry. The three Hammerwood localities, a forge and possibly a furnace at Cansiron [about 453 382] are sited on Ashdown Beds some distance from the nearest Wadhurst Clay outcrop.

Large slabs of 'Cyrena' limestone approximately 20 ft (6·1 m) from the top of the Wadhurst Clay were found in a ditch [4134 3789] 700 yd (640 m) W. of Shovelstrode Manor. The limestone is 1 to 2 in (25–51 mm) thick and crowded with valves of

'*Cyrena*', mostly crushed, although weathered, better preserved, specimens occur on the upper surface.

A bed of sand and sandstone, over 18 ft (5·5 m) thick was proved in a borehole at Ivy Dene Laundry, Ashurstwood [4168 3703]. The complete record beneath 19 ft (5·8 m) of Made Ground and Lower Tunbridge Wells Sand in downward sequence is: blue clay, 7 ft (2·1 m); sandstone, 1 ft (0·3 m); blue clay, 23 ft (7·0 m); blue clay and sandstone, 17 ft (5·2 m); sandy blue clay, 5 ft (1·5 m); sandy black slatey clay, 10 ft (3·0 m); sand and sandstone, 10 ft (3·0 m); fine brown blowing sand with sandstone layers, 8 ft (2·4 m); total depth of the hole, 100 ft (30·5 m). No evidence of this sandstone could be found at outcrop.

The reddened top of the Wadhurst Clay was augered in Chance Coppice [4102 3661] 400 yd (366 m) N W of Brambletye. Red Clay was also found at a point [4545 3624] 700 yd (640 m) at 250° from St. Ives Farm, but within a few feet of the base of the Wadhurst Clay.

Ironstone nodules were dug from the ditch [4592 3742] 300 yd (274 m) at 060° from Beeches Farm, which is 10 to 15 ft (3·0–4·6 m) above the base of the Wadhurst Clay. A well [4712 3663] 400 yd (366 m) at 260° from Hartwell penetrated 47 ft (14·3 m) of Wadhurst Clay beneath the Lower Tunbridge Wells Sand (Harvey and others 1964, 303/106). Ironstone was encountered in the top 17 ft (5·2 m) of Wadhurst Clay. Straker (1931, pp. 241–5) found evidence for a furnace and a forge at Parrock [4575 3567] and also near Chartness (Chartner's) Farm [about 475 363], sited on the flood plain of the River Medway. C.R.B.

Charlwood Graben. The Wadhurst Clay of the Charlwood Graben was regarded as clay within the Ashdown Beds by Topley (1875, p. 80), although part of the structure around Claypits Farm (now Tylehurst [4120 3405]) was mapped as Wadhurst Clay.

Poor exposures of Wadhurst Clay occur in some of the stream courses within this fault complex. Sections at 400 yd (366 m) at 055° from Charlwood [3962 3444] and at 600 yd (549 m) at 105° from Charlwood [3987 3407], show deeply weathered valley-bulged grey shales overlain by thick sandy hillwash deposits. Similar valley-bulged shales containing '*Cyrena*' outcrop in the stream bed [4078 3404] 350 yd (320 m) at 110° from Mudbrooks Farm.

Numerous deep pits occur within the complex, notably adjacent to Charlwood [392 341] and south-west of Mudbrooks Farm [402 340]. These were doubtless worked for 'marl' but may also have supplied some ironstone to the bloomeries at Whalesbeech and Stonefield [397 342] (Straker 1931, p. 240). Several of these workings expose the red clays which mark the top of the formation. R.W.G., C.R.B.

Hartfield. There are now no exposures of clay in the Upper Hartfield outlier although numerous flooded pits bear witness to former extensive working. 'Blue' clay was worked in the brickyard [451 342] just north of Coleman's Hatch church until recent years (about 1960).

Possibly ironstone from this outlier was worked at the bloomery at Cotchford (see also p. 64) and the furnaces at Cotchford and Newbridge (Straker 1931, pp. 248–52).

Boarshead. To the south of the Silverlands Fault the Wadhurst Clay dips 5° at 010°. The clay is overlain by Lower Tunbridge Wells Sand and the outcrop dissected by numerous streams. The thickness of the Wadhurst Clay is about 120 ft (36·6 m), although cambering may have attenuated the outcrop (see p. 112).

Red clay at the top of the Wadhurst Clay was proved in the fields [5276 3274 and 5407 3270] ½ mile (0·8 km) W. and 600 yd (548·60 m) E. of Boarshead, and in the road bank [5496 3268] 650 yd (594 m) S.E. of Sand Hill Farm.

The presence of former iron workings is evidenced by iron slag in the stream [5257 3237] 700 yd (640 m) at 012° from Hourne Farm. Straker (1931, p. 252) recorded bloomeries at Grubs [about 500 329] and Steel Cross [about 529 318] and a furnace ½ mile (0·8 km) W. of St. John's Church, Crowborough.

On the western side of the angle in the Crowborough–Tunbridge Wells road at Steel Cross are numerous old diggings, several acres across, in the basal Wadhurst Clay. It was mostly marl which was extracted, but the licence to dig the marl included the right to sell the iron ore which is common at this horizon (Straker 1931, p. 263). To the north of the Silverland Fault bell-pits are to be found in Minepit Wood [5217 3390 and 5236 3395], 500 yd (457 m) N. of Orznash. Close to these pits is a large dump of iron slag [5232 3382] and it is here that two bloomeries are located (Straker 1931, p. 257). Excavations for the Sussex Archaeological Society in 1965 and 1966 revealed a well-preserved furnace [5237 3393], the age of which is uncertain; an Iron Age–Romano-British date, based on pottery evidence, has been suggested, but structurally it appears to be Saxo-Norman to mediaeval. A bloomery, a forge and a furnace formerly existed at Furnace Fields, ¾ mile (1·2 km) W. of Eridge Station and at Birchden and Hamsell (Straker 1931, pp. 260–2).

A trench section [5280 3445] 200 yd (183 m) at 160° from Bullfinches showed that the uppermost Wadhurst Clay was reddened.

Tunbridge Wells–Lamberhurst. A thickness of 178½ ft (54·4 m) for the Wadhurst Clay was proved at Kelsey's Culverden Brewery, Tunbridge Wells (Whitaker 1908, pp. 209, 381; Harvey and others 1964, 303/78), while just to the north of the present area at Bassett's Farm [6411 4179] a well sited just below the base of the Lower Tunbridge Wells Sand proved 190 ft (57·9 m) of Wadhurst Clay beneath 11 ft (3·4 m) of Head (Whitaker 1908, p. 179).

At High Brooms red clay at the top of the Wadhurst Clay was proved in a temporary excavation [5921 4119] and by augering [5951 4115] (see also Dines and others 1969). East of Broomhill, on the northern margin of the map, the Wadhurst Clay outcrop is hidden by superficial downwash from springs issuing from the base of the Tunbridge Wells Sand. Temporary sections for the new housing estate to the west of Gregg's Wood proved that the sandy wash was at least 5 ft (1·5 m) thick; at one point [6009 4080] 5 ft (1·5 m) of it overlie the reddened top of the Wadhurst Clay. Farther east [640 412] it was not possible to auger through the wash to the clay beneath, although downstream, on the adjacent Sevenoaks (287) Sheet, clay was proved at several localities. The springs issuing from the base of the Tunbridge Wells Sand feed the Pembury Reservoir.

'*Cyrena*' limestone was revealed in stream sections [6212 3976, 6302 3899, 6309 3919, 6326 3940 and 6481 3860] although most of the outcrops are disturbed by valley bulging. At one point [6309 3919], 750 yd (686 m) at 105° from Great Bayhall, the limestone is 4 in (0·1 m) thick. Associated with limestone in this stream section are exposures of silts and siltstone [6310 3925, 6325 3930, 6398 3943, 6326 3940] and ironstone [6296 3875, 6327 3940 and 6355 3956]. Ironstone nodules have also been found in the stream [6212 3975] 500 yd (457 m) N.W. of Great Bayhall, and have been worked in bell-pits [6260 3835 and 6350 3815] in the woods to the east and west of Dundle. There is much iron slag in the shallow valley [6123 3847] 1000 yd (914 m) at 015° from Benhall. Straker (1931, pp. 264–7) recorded forges at Benhall, Melhill [about 6130 3905], Breecher's Forge or Marriot Croft [about 627 385] and Dundle.

The reddened top of the Wadhurst Clay was augered at several points along the outcrop [6166 3850, 6178 3853, 6186 3861, 6292 3976, 6364 3989, 6367 3982, 6378 3982 and 6362 3844].

C.R.B.

In its extensive outcrop east of Lamberhurst Quarter the Wadhurst Clay is about 160 ft (48·8 m) thick. Exposures are few and poor at the present day, but the many old flooded pits which dot its outcrop testify to the widespread working of the clay and contained ironstone in the past. Numerous old bell-pits occur in Clayhill Wood [653 377] marking the site of ironstone workings in the basal beds of the formation.

A recent temporary excavation [6714 3713] 160 yd (146 m) at 280° from Windmill Farm exposed the contact with the Tunbridge Wells Sand:

	Ft	in	(m)
Tunbridge Wells Sand:			
Fine ochreous siltstone with thin sandstone, rather obscured	2	0	(0·61)
Wadhurst Clay:			
Reddened ochreous clay	2	0	(0·61)
Ochreous clay ironstone	0	1	(0·025)
Thin grey cross-laminated siltstones with traces of rootlets; hard indurated 2-in (50-mm) siltstone at top..	0	8	(0·20)
Bluish grey clay	1	6	(0·46)
Greyish purple silty clay, full of lignitic plant debris and with small ochreous clay ironstone nodules up to 1 in (25 mm) in diameter	1	0	(0·30)
Clay ironstone, traces of vertical roots	0	3	(0·076)
Cross-bedded, calcareous siltstone, with plant debris	0	1	(0·025)
Ochreous greyish green clay with thin siltstones, up to ..	2	6	(0·76)

The occurrence of ironstone in these upper beds is of interest, possibly having had some influence in the siting of old pits along the Wadhurst Clay–Tunbridge Wells Sand junction in this area.

Numerous small exposures occur in the bed of the stream rising at Lamberhurst Quarter and thence coursing east-north-east towards Tong Farm [6748 3973]. In most cases the shales and siltstones exposed are disturbed and contorted by valley bulging. In the south bank of the stream [6722 3964] 300 yd (274 m) at 250° from Tong Farm up to 5 ft (1·5 m) of bluish grey shales interlaminated with thin siltstones and containing ostracods and *Neomiodon* shells crop out; fragments of a ¾-in (19-mm) '*Cyrena*' limestone occur in the stream bed. Farther upstream [6630 3914], 570 yd (521 m) at 320° of Little Dunks Farm, up to 3 ft (0·9 m) of greyish green shales and siltstones with a 4-in (0·1-m) ferruginous '*Cyrena*' limestone, inclined sharply to the north-east, are exposed in the south bank.   E.R.S.-T.

To the east of the Coneyburrow Wood Fault, Tunbridge Wells Sand and Wadhurst Clay are let down between the North and South Groombridge faults. The Wadhurst Clay outcrop is limited and partially obscured by Head or Alluvium. Around Bayham Abbey [645 368] the thickness of the Wadhurst Clay appears to be only about 80 ft (24·4 m); this is probably an effect of cambering. The reddened top of the clay was found in the field [6470 3675] east of the Abbey and in the small inlier brought to the surface by valley bulging in the stream bed [6500 3749] ½ mile (0·8 km) N.E. of Bayham Abbey (see p. 111).

There is now little evidence of iron workings on the Wadhurst Clay but Straker (1931, p. 268) recorded a furnace at Tollslye [632 371] to the south of Furnace Wood; another at Furnace Mill [662 332], Lamberhurst; and a forge at Bayham Abbey [about 642 366] to the south-east of Forge Wood.   C.R.B.

Wadhurst Clay crops out on the northern flanks of the Teise Valley, north of the South Groombridge Fault, between Bayham Abbey and Lamberhurst; its thickness in this region is apparently less than 150 ft (45·7 m), though the true thickness is probably greater.

There are no present-day exposures of the formation in this tract, although there are numerous old diggings. These include large pits near the base of the formation at Deep Shaw [651 368] and Minepit Shaw [6545 3673], presumably worked for ironstones, and some just below its junction with the Tunbridge Wells Sands, notably one at the south-east corner of Timberlog Wood [6664 3638]. Reddened clays are a feature of the topmost Wadhurst Clay throughout this area.

Steep slopes cut in the Wadhurst Clay are prone to landslipping and two minor examples occur hereabouts at one location [658 371], 350 yd (320 m) at 020° from Hoathly Cottages, and at another [664 365] in Timberlog Wood.

Recent water bores at Furnace Farm (Mid-Kent Water Company) have proved the presence of Wadhurst Clay beneath the Tunbridge Wells Sand that crops out between the Groombridge and Kilndown faults. The results, based on examination of samples taken during percussion drilling, are summarized below:

Borehole	No.1 Trial	No. 1	No. 2
Surface Level	143·6 ft	152·3 ft	145·1 ft
	(43·8 m) O.D.	(46·4 m) O.D.	(44·2 m) O.D.
National Grid Ref.	[6667 3605]	[6669 3600]	[6654 3596]
Tunbridge Wells Sand	c.30 ft (9·1 m)	c.30 ft (9·1 m)	89 ft (27·1 m)
Wadhurst Clay	c.215 ft (65·5 m)	c.190 ft (57·9 m)	340 ft (103·6 m)
		up to	proved
Ashdown Beds	c.105 ft (32·0 m)	c.45 ft (13·7 m)	

The abnormally large thickness of Wadhurst Clay proved in the No. 2 Borehole probably has a structural origin due to the proximity of the Kilndown Fault. There is no surface evidence of reversed faulting in the area, and a likely explanation would be that the borehole has penetrated Wadhurst Clay in a fault slice or which alternatively is steeply dipping owing to terminal bending of strata towards the fault.   E.R.S.-T.

Eridge-Frant. At Frant the Wadhurst Clay is about 200 ft (61·0 m) thick. Two boreholes situated close to the crest of the Frant ridge proved thicknesses of 197 and 215 ft (60·0, 65·5 m) (see Harvey and others 1964, 303/35, 303/130c).

Flaggy ironstone, in the upper part of the Wadhurst Clay, is exposed in the stream [6065 3509] 300 yd (274 m) at 250° from Bareland. Farther downstream [6065 3504] flaggy siltstone is to be seen.

In a stream section in Clay's Wood [6047 3495], about 1 mile (1·6 km) at 105° from Frant, were 6 ft (1·8 m) of thinly interbedded medium grey mudstones and ripple-marked siltstones. *Equisetites lyelli* was found in the form of flattened carbonaceous rhizomes, with associated vertical roots and rootlets indicating the presence of a soil bed 'in situ'. Valley bulging, typical of Wadhurst Clay stream exposures, has caused the rocks to be sharply folded (Plate VIIIA).   R.A.B.

At the top of the Wadhurst Clay red clay can be found continuously around the Eridge Castle outcrop, and at several other localities farther east [5746 3588, 5741 3542, 5862 3604, 5914 3625, 5920 3507 and 5922 3532].

A forge (Old Forge [5597 3507]) and a furnace [about 5635 3507] were formerly located in Eridge Park (Straker 1931, pp. 257–8).   C.R.B.

Site investigation boreholes at Bartley Mill (see p. 84 and Appendix I) [631 355], 800 yd (732 m) at 196° from Wickhurst, proved up to 45 ft (13·7 m) of olive grey, dark grey and red mottled mudstones of the topmost Wadhurst Clay. Red staining was most prominent within the topmost 15 to 20 ft (4·6–6·1 m) adjacent to the junction with the Lower Tunbridge Wells Sand. Beds of hard siltstone, up to 1 in (25 mm) thick and containing casts of rootlets and vertical stems up to $\frac{1}{10}$ in (3 mm) in diameter occur at several levels.   R.W.G.

Mark Cross area. In Marchant's Wood, 400 yd (366 m) at 040° from Redgate Mill Farm, there are signs of past ironstone mining in the form of bell-pits. Similar workings were found 280 yd (256 m) at 260° from Saxonbury Hill Tower. In both cases the bell-pits were in the lower 30 ft (9·1 m) of the Wadhurst Clay. At the latter locality the abrupt limit of bell-pitting is an indication of the termination of the ironstone beds against the Withyham Fault.

In a small stream 200 yd (183 m) N. of Fright Farm (now Saxonbury Farm) are contorted medium grey mudstones. The contortions may be partly connected with the nearby fault but are probably mainly due to valley bulging (see p. 112). A section [5871 3255] estimated to be 20 ft (6·0 m) above the base of the Wadhurst Clay showed, in downward sequence: medium grey mudstones, 2 ft 7 in (0·8 m); reddened clay ironstone nodules, 2 in (50 mm); medium grey mudstones, 6 in (152 mm); reddened limestone with *Neomiodon*, 2 in (50 mm); medium grey mudstones, 7 in (178 mm); medium grey siltstones and mudstones, finely interbedded, 2 ft (0·6 m).

Mark Cross [5825 3115] was once a major centre in the ironstone industry. Bell-pits are found all round the village, at horizons no more than 40 ft (12·2 m) above the base of the Wadhurst Clay. They were noted just below Catts Farm [5810 3140] and the thin seams of ironstone were probably stripped from the ground where near the surface just north of Mark Cross [5830 3170]. Many old bell-pits still exist in the woods north of Mark Cross, particularly in Frankham Wood [5870 3200] and Pond-field Shaw [5960 3200]. Bloomeries have been found at Colegrove [about 590 330], Earlye Farm [5975 3290], Sandyden [586 309], Stilehouse Wood [5850 3030] and Chant Farm [about 559 304] and furnaces at Cowford [about 560 319], Henly [about 601 337] and Riverhall [about 605 333] (Straker 1931, pp. 256, 273–6, 288; Cattell 1970).

Some 100 yd (91 m) W. of Frankham a small temporary section [5893 3153] revealed about 4 ft (1·2 m) of ripple-marked siltstones with load casts, thinly interbedded with muds. Rootlets of *Equisetites lyelli* were found in some of the siltstones.   R.A.B.

Bewl and Hook River valleys. The Wadhurst Clay floors the valleys of the River Bewl and its tributary the Hook River, south of the Chingley Fault, and extends up the valley sides to heights between 200 and 400 ft (61–121·9 m) above O.D. The valley is the site of a proposed reservoir and the dam site has been explored by numerous trial borings (1961–2) which give useful lithological data on the formation (Shephard-Thorn and others 1966, p. 49). Natural exposures are very few in the valleys; numerous old pits occur along the junction with the Tunbridge Wells Sand, which is often marked by the occurrence of red clays.

Shover's Green–Ticehurst. The flat-topped west to east ridge between Shover's Green and Ticehurst is capped by Wadhurst Clay, a thickness of up to 100 ft (30·5 m) being preserved. The formation also crops out in the complex zone of faults between Broomden [6780 3070] and Rowley [6790 3160] but there are no exposures.   E.R.S.-T.

Danehill–High Hurstwood–Buxted. The thickness of the clay was 176 ft (53·6 m) in a borehole at Dallings [4835 2497], 300 yd (274 m) N.W. of Bevingford, and 147 ft (44·8 m) at Warminghurst [493 248], ½ mile (0·8 km) E. of Bevingford (Edmunds 1928, pp. 69, 70; Harvey and others 1964, 303/17, 303/16). The lower part of the Wadhurst Clay may have been cut out by the Hendall Wood Fault in the Warminghurst Borehole. A borehole at Moaps [4009 2699], 550 yd (503 m) at 195° from Danehill church, which commenced about 10 ft (3·01 m) above the base of the Lower Tunbridge Wells Sand, proved Wadhurst Clay consisting of dark shales with thin silt and sandstone beds to 195 ft (59·4 m). The base of the formation was not reached.

    C.R.B., R.W.G.

At Westlands [3910 3058] the basal 10 ft (3·0 m) of the formation are very silty, the junction with the Ashdown Beds being marked by a poorly developed pebble bed. An old shallow pit there worked the basal ironstone.

A section in a road cutting [3982 2874] 450 yd (411 m) S. of Warren House shows cream-coloured silts resting on a ripple-marked surface of coarse ferruginously cemented sandstone, which in turn rests on flaggy fine-grained sandstones. The relationship of this small outlier to similar exposures on the Horsham (302) Sheet suggests that these silts are at the base of the Wadhurst Clay and not part of a cyclothem within the Ashdown Beds.

F

Stream sections [398 273] in the small inlier 1100 yd (1 km) at 260° from Danehill church show valley-bulged grey shales overlain by thick sandy hillwash deposits.

In the fault-bounded outlier of Wadhurst Clay immediately north of Danehurst an almost continuous line of deep pits in red and grey clays, extending from Cowstocks Wood [393 268] through Pitt Wood [397 269] to Upper Dane Wood [405 270], marks the top of the formation.   R.W.G.

Bell-pits are present in Dane Wood [4075 2710], 800 yd (805 m) S.E. of Danehill church. A tip of iron-slag at Sheffield Mill [4153 2573], 800 yd (805 m) at 120° from Furner's (i.e. Furnace) Green, indicates the presence of the old furnace sited here (Straker 1931, pp. 412–4).

A small inlier, fault bounded to the north, occupies the bottom of the valley [436 263] east of Woolpack Farm.

Around Forest Lodge and Fairwarp there is a larger fault-bounded outcrop. At both localities there are numerous pits, many flooded, but no exposures at the present day. Bell-pits can be found 600 yd (549 m) E. of Forest Lodge [4578 2615] and 500 yd (457 m) at 075° from Cophall Farm [4703 2637]. They were also noted farther east in Minepit Wood [4805 2695] near Heron's Ghyll and in Parkhurst Wood [4887 2595], 550 yd (503 m) N. of Stonehouse. At the last three localities the bell-pits are within 20 ft (6·1 m) of the base of the Wadhurst Clay. There is much iron clinker in Furnace Wood [4775 2660] to the north-east of Claygate Farm. The iron in this district was worked by the Romans just north of Furnace Wood [about 476 268] and the cinder from this site was being used for road metal in 1844 (Straker 1931, pp. 395–7). In Tudor times, a furnace was established farther upstream [about 4765 2715] and is thought by Straker (1931, pp. 387, 394) to be the furnace where the first Sussex iron guns were cast in 1543. Furnaces have also been found at Hendall (Cinder Bank [470 250]) and at Old Forge [4586 2577], where, as its name implies, there was also a forge. Boringwheel Mill Farm [456 264], Cackle Street, is thought to have been a forge (Straker 1931, pp. 394–9).

Away from the Wadhurst Clay outcrops forges are recorded by Straker (1931, pp. 400–3, 412–5) at Sheffield Bridge, Fletching, and in Maresfield Park (Forge Wood, 452 239). There is much iron slag to the north of Forge Wood [4527 2417]. Lower down this same stream (Batts Brook) was sited a forge and a furnace [about 422 230] (Straker 1931, pp. 400–3).

Good exposures of grey shaly clay can be found in the small stream (Claygate Gill [4815 2610]) to the east of Claygate Farm. Limestone was noted 500 yd (457 m) at 160° from the farm [4803 2584], and 'Cyrena' limestone on the west bank [4932 2563] of the stream which flows through High Hurstwood.

Red clay was frequently encountered just beneath the Tunbridge Wells Sand [4773 2640, 4798 2670]; continuously along the east part of the outcrop ½ mile (0·8 km) E.S.E. from Heron's Ghyll, and also in the northern part of the Stonehouse outlier. More unusually, red clay was found some 25 to 30 ft (7·6–9·1 m) from the top of the Wadhurst Clay and overlain by grey clay [4775 2660, 491 248, 4968 2495, 4954 2474, 4952 2398, 4910 2343 and 4934 2285]. Red clay beneath 3 ft (0·9 m) of grey clay at the top of the Wadhurst Clay has also been proved in a borehole at Dallings [4835 2497] 300 yd (274 m) N.W. of Bevingford (Edmunds 1928, p. 69; Harvey and others 1964, 303/17).
                                                                          C.R.B.

Steep Park–Argos Hill area. In a deeply incised stream 170 yd (155 m) at 290° from Rumsden Farm [5284 2769] a section shows: medium grey shaly mudstone, 1 ft (0·3 m); shell bed with *Neomiodon*, ½ in (13 mm); medium grey shaly mudstone, 2 ft 4 in (0·7 m); nodular clay ironstone, ½ in (13 mm); medium grey mudstone grading down into mudstone interlaminated with siltstones, 3 ft 4 in (1·0 m); medium grey shaly mudstone, 1 ft (0·3 m).

This sequence is probably about 50 ft (15·2 m) above the base of the Wadhurst Clay. An old bell-pit above the section indicates that ironstone was once mined at this locality.

About 400 yd (366 m) S. of Castle Hill a temporary section [5520 2764] exposed 6 ft (1·8 m) of medium grey shaly mudstones with thin siltstone layers; it included a nodular clay ironstone just under 1 in (25 mm) in thickness and a limestone with *Neomiodon*. The grey mudstones contained *Neomiodon* and *Viviparus*. The level is estimated to be about 80 ft (24·4 m) above the Ashdown Beds.

A section at the base of the Wadhurst Clay 780 yd (713 m) at 130° from Rother-hurst (formerly Churt House) in the bank of the River Rother, showed just over 2 ft (0·6 m) of medium grey shaly mudstones on massive sandstones and silts. There was no sign of the Top Ashdown Pebble Bed at this locality.

The top strata of the Wadhurst Clay were exposed as follows in a railway cutting [5758 2822] just 160 yd (146 m) at 329° from Park Farm: Tunbridge Wells Sand comprising light coloured hard siltstone, 1 ft (0·3 m); light grey flaggy silty sandstones, 3 ft 3 in (1·0 m); Wadhurst Clay comprising red clay, 1 ft (0·3 m) and medium and light grey shales, in places mottled red and grey bands, poorly exposed to 10 ft (3·0 m).

Rarely are large thicknesses of Wadhurst Clay exposed in this area, but several sections in a stream through the southern part of Spitlye Wood [5760 2870], about 600 yd (549 m) at 294° from Great Wallis Farm, enabled the following composite sequence to be built up:

	Ft	in	(m)
Argillaceous limestone; lower surface with load casts; *Neomiodon*	0	1½	(0·04)
Medium grey shaly mudstone	1	7	(0·48)
Argillaceous limestone with *Neomiodon*	0	1	(0·025)
Medium grey shaly mudstone	1	5	(0·43)
Medium grey shaly mudstone with *Neomiodon* shell bands ..	1	6	(0·46)
Medium grey shaly mudstone	1	7	(0·48)
Medium grey mudstone and nodular limestone with *Neomiodon*	0	0½	(0·01)
Estimated gap to next section	6	0	(1·82)
Medium grey mudstones	3	0	(0·91)
Shelly limestone; variable thickness .. .. ..2 in to	0	8	(0·05– 0·20)
Medium grey shaly mudstone with *Neomiodon*	6	0	(1·83)
Argillaceous limestone with *Neomiodon*	0	2	(0·05)
Estimated gap to next section	13	0	(3·96)
Medium grey shaly mudstone with *Neomiodon* shell layers and thin shelly limestones	2	8½	(0·82)
Estimated gap to next section	3	0	(0·91)
Medium grey shaly mudstone	1	0	(0·30)
Medium grey shaly mudstone with thin argillaceous limestones and bands rich in *Viviparus* and *Neomiodon*	2	0	(0·61)
Medium grey mudstone	1	6	(0·46)
Medium grey shaly mudstones and limestones with *Neomiodon*; limestones 1 to 2 in thick; load casts and ripple marks ..	2	0	(0·61)
Estimated gap to next section	8	0	(2·44)
Medium grey shaly mudstone with soil bed; *Equisetites lyelli* rhizomes and rootlets; ostracods	1	6	(0·46)
Blocky medium grey mudstones; *Neomiodon*, *Equisetites sp.*	1	0	(0·30)
Estimated gap to massive sandstones of the Ashdown Beds ..	27	0	(8·23)

The above stream section shows parts of the basal 85 ft (25·9 m) of the Wadhurst Clay succession. Notable is the absence of clay ironstones which are probably present in the unexposed parts of the sequence and the presence of an 8-in (203-mm) shelly limestone.

Rushers Cross–Wadhurst Hall–Stonegate. In Butcher's Wood [6191 2928] some 6 ft (1·8 m) of thinly bedded medium grey shale and silts were seen in a stream section. Farther up the stream are many small exposures of shales and silts with thin bands rich in *Neomiodon*. These sections are about 120 ft (36·6 m) below the top of the Wadhurst Clay.

Near Wadhurst Park Farm (formerly Wadhurst Hall [6322 2888]) three boreholes made in 1960 gave a virtually complete succession through the Wadhurst Clay (Anderson and others 1967). Although two thin fine-grained sandstones were recorded in these boreholes only the lower of these, about 5½ ft (1·7 m) thick, could be traced at the surface, and this only for a short distance.

Some 400 yd (366 m) at 045° from Bricklehurst Manor, near Stonegate, a small knoll of sandstone in the Wadhurst Clay was found by augering [6623 2955]. This is only 100 yd (91 m) N. of the Limden Fault and is probably in the lower part of the Wadhurst Clay succession.

In the stream [6595 2806] at Hoadley Wood, about 20 ft (6·1 m) above the base of the Wadhurst Clay, there were several small exposures. The estimated succession in downward sequence is: medium grey shale, 1½ ft (0·5 m); hard light grey (almost white) silt, 1 ft (0·3 m); medium grey silts and shales with cross-bedded and ripple-marked silts at the base, becoming more shaly upwards, 3½ ft (1·1 m); yellowish brown silt, 1 ft (0·3 m).

The above section indicates the silty nature of the Wadhurst Clay, particularly near the base. In the stream through Cock's Wood [6661 2689], 600 yd (549 m) at 290° from Hammerden, silty beds were again exposed at similar horizons.   R.A.B.

An area of dull red and chocolate-coloured clay adjacent to the Limden Fault [677 292], some 400 yd (366 m) W. of Wedd's Farm, may be near the top of the formation.

Around Battenhurst Farm the Wadhurst Clay dips south at 1° to 2°. Sections very close to the base of the formation, in a road cutting [6788 2693] 300 yd (274 m) S.E. of Battenhurst Farm, show siltstones containing a variety of trace fossils overlain by clays with included beds of clay ironstone and a 3-in (76-mm) thick '*Cyrena*' limestone.

Old bell-pits for ironstone are abundant in the lower 30 ft (9·1 m) of the Wadhurst Clay from Ten Acre Wood [6315 2715], 800 yd (731 m) at 50° from Hareholt Farm, to Newbridge Wood [6400 2690]. Minepits Shaw [6070 2800], 400 yd (366 m) at 330° from Sharnden Farm, and Cinderhill are indicative of the old industry. Clay ironstone debris is common in the fields adjacent to the junction with the Ashdown Beds, and many of the old pits between Battenhurst Farm and Cottenden probably worked this horizon. Bell-pits in the woodland [6808 2810] 600 yd (549 m) S.E. of Cottenden are at the same horizon. Cattell (1970) has found 13 bloomery sites in the triangular area—Little Trodgers [590 300]–Cinderhill [602 283]–Pennybridge [612 308] additional to those listed by Straker, who noted them nearby at Bardown (Roman) [6605 2942], Hammerden [6573 2700] and Shoyswell [6871 2731], and a furnace at East Limden [6772 2904]. Excavations at Bardown by the Wealden Iron Research Group during 1967–8 indicate that industrial activity, established about A.D. 140–150, ceased at the end of the 2nd century. Occupation of the site continued and satellite working sites were set up about 1 mile (1·6 km) to the north. It is estimated that 10 000 tons of iron were made at Bardown (Cleere 1970).   R.A.B., R.W.G.

Hastingford, Five Ashes area. North-west of Hadlow Down, 150 yd (137 m) at 330° from Gate House, sandstone at an horizon about 40 ft (12·2 m) above the base of the Wadhurst Clay was traced around and on three ridges just north of the Greenhurst Fault.

Bell-pits are abundant in Oven's Mouth Wood and in Stockland Wood about 500 yd (457 m) N. of Stockland Farm. Ironstone was extensively dug from the lower beds of the Wadhurst Clay all around this inlier of the Ashdown Beds.

In Vicar's Wood, 400 yd (366 m) at 202° from Skipper's Hill were about 2 ft (0·6 m) of medium grey shale with 1-in (25-mm) limestones containing *Neomiodon* and *Viviparus*. The relatively uncommon occurrence of *Viviparus* is referred to elsewhere (Anderson and others 1967, p. 176).

Good sections of medium grey mudstones with thin nodular ironstones and limestones with *Neomiodon*, all highly contorted owing to valley bulging, occur in the stream through Almond's Wood [5662 2515] at about 870 yd (796 m) at 128° from Butcher's Cross.

At 860 yd (786 m) 125° from Five Ashes, in the stream through Roundbacks Shaw, there is a series of excellent exposures in the upper beds of the Wadhurst Clay. The estimated sequence of strata, with grid references given for each individual section, is:

	Thickness		
	Ft	in	(m)
[5643 2436]			
Medium grey mudstone grading down into cross-bedded siltstone; some bivalve shells	2	8	(0·81)
As above but siltstone less thick; ostracods	2	0	(0·61)
Medium grey mudstones	0	6	(0·15)
Shell bed with *Neomiodon*	0	0¼	(0·006)
Medium grey mudstone with soil bed; *Equisetites* stems and rhizomes all through	2	0	(0·61)
Estimated section obscured	5	0	(1·52)
[5645 2440]			
Medium grey mudstones	1	10	(0·56)
Sandy limestone with cone-in-cone structure above upper surface	0	3	(0·076)
Medium grey mudstones and siltstone	1	11	(0·58)
Estimated section obscured	2	0	(0·61)
[5650 2446]			
Medium grey mudstones	6	0	(1·83)
Shelly limestone with load casts at base	0	2	(0·05)
Estimated section obscured	2	0	(0·61)
[5654 2448]			
Yellow weathering siltstones	2	0	(0·61)
Medium grey mudstones	3	8	(1·2)
Nodular clay ironstone	0	1	(0·025)
Medium grey mudstones and siltstones interbedded; soil bed of *Equisetites*, vertical roots common	1	3	(0·38)
Estimated section obscured	10	0	(3·05)
[5667 2456]			
Grey thinly bedded siltstones with shales	1	1	(0·33)
Medium grey mudstones with silts; *Neomiodon*	1	4	(0·41)
Grey siltstones; vertical roots at top ? *Equisetites*	0	11	(0·28)
Shell bed with *Neomiodon*	0	0½	(0·012)
Medium grey mudstones with *Equisetites* rhizomes and stems	0	6	(0·15)
Medium grey shelly mudstones	1	1	(0·33)
Shell bed with *Neomiodon*	0	0¼	(0·006)
Medium grey mudstone	0	7	(0·18)

	Thickness
	Ft in (m)
Calcareous siltstone filling channels in mud below	0 0½ (0·012)
Medium grey mudstones with thin siltstones	2 6 (0·76)

That these strata are in the upper part of the Wadhurst Clay is confirmed by the ostracod assemblages.

Mayfield–Burwash. Near Mayfield the Wadhurst Clay is brought up by sharp folding near the South Mayfield Fault [590 268] to form a small sinuous inlier detected by augering. To the south of this, 360 yd (329 m) at 081° from Cranesden, about 4 ft (1·2 m) of massive ripple-marked sandstone are exposed in a small deep-cut stream [5893 2633]. This sandstone, about 50 ft (15·2 m) below the top of the Wadhurst Clay, is weathered to a light brown colour and contains casts of bivalve shells; medium grey mudstones above and below are strongly affected by valley bulging. The sandstone could only be traced for a short distance.

Farther to the east, 300 yd (274 m) N. of Hareholt Farm, a small strip of Wadhurst Clay crops out as a result of folding near the North Mayfield Fault. The red clays typical of the upper part of the Wadhurst Clay are evident hereabouts.

To the south of the last locality, about 500 yd (457 m) at 041° from Bignowle, a stream [6258 2478] exposed in downward sequence: Wadhurst Clay comprising medium grey mudstone (weathered to clay), 2 ft (0·6 m); oolitic ironstone, 2 in (50 mm); pebble bed becoming coarser downwards and with ironstone nodules along the base, 2 in (50 mm); Ashdown Beds consisting of sands and silts, weathering light brown, 2 ft (0·6 m). The oolite here is unusual for the area.

Some 1500 yd (1·4 km) S. of Burwash church a section was seen in 1949 in the south corner of an old clay pit [6774 2336]; before it became overgrown the following approximate measurements were made by Mr. B. C. Worssam:

	Thickness
	Ft in (m)
Wadhurst Clay:	
Clay and shale with a thin irony shell bed with small bivalves near the top; rhizomes of *Equisetites lyelli* occur in rusty mottled light grey clay about 7 ft (2·1 m) from the base..	15 0 (4·6)
Alternating brown shales and flaggy siltstone	1 6 (0·5)
Pebble bed, coarse gritty loosely coherent sand with layer of pebbles, 1 in (25 mm) thick, at base; sharp junction with underlying bed	0 7 (0·2)
Ashdown Beds:	
Fine-grained, honey-coloured sandstone	6 0 (1·8)

The soil bed in the above section is the Brede *Equisetites lyelli* Soil Bed (Allen 1947).

Isenhurst Park–Broadoak–Burwash Weald. A stream section [5716 2385] some 940 yd (860 m) at 67° from Button's Farm, south of Five Ashes, exposed the following descending sequence, strongly folded owing to the proximity of a fault (the folding possibly also accentuated by later valley bulging): medium grey mudstones, 2 ft (0·6 m); shelly limestone, ½ in (13 mm); medium grey mudstone, 1 in (25 mm); clay ironstone, 2 in (50 mm); medium grey mudstone, 3 in (76 mm); yellow silty mudstone with shells, 1 in (50 mm); medium grey mudstone with shells, 7 in (178 mm); nodular clay ironstone, 1 in (25 mm); medium grey mudstone, 2 ft (0·6 m). This section is probably in the lower part of the Wadhurst Clay.

In the two outliers of Wadhurst Clay about 1 mile (1·6 km) N.W. of Broadoak only the lower half of the Wadhurst Clay is present. In both outliers bell-pits for the basal ironstones are very common, i.e. in Oaken Wood [5885 2385] and near the Pheasantry [5962 2370].

In a small faulted outlier just south of the Burwash Common Fault, 370 yd (338 m) at 262° from Green Hill, Burwash Weald, the following basal Wadhurst Clay sequence was exposed [6505 2350]: Wadhurst Clay consisting of medium grey mudstones with ironstone nodules 0 to 1 in (0–25 mm) thick, 1 ft 6 in (457 mm); fine-grained yellow ripple-marked sandstone, ½ in (13 mm); pebble bed (variable thickness) about 1 in (25 mm); Ashdown Beds comprising fine-grained light grey sand, 6 in (152 mm); massive yellow sandstone, 4 in (102 mm). R.A.B.

LOWER TUNBRIDGE WELLS SAND

Dormans Park–Kent Water Valley–Holtye Common. Much of this outcrop was formerly shown as Ashdown Beds, the Grinstead Clay having been mistaken for Wadhurst Clay. The correct sequence was recognized during the six-inch survey of the Sevenoaks (287) Sheet in 1933–5. West of Basing Farm the outcrop is broad and little affected by faulting, but to the east it is complicated by faulting; the whole is faulted down to the north against Ashdown Beds by the Blockfield Fault.

The thickness of the Lower Tunbridge Wells Sand is about 100 ft (30·5 m), of which 40 ft (12·2 m) is assigned to the Ardingly Sandstone. To the south of the Kent Water the Lower Tunbridge Wells Sand has a narrow outcrop. The lower part of the succession is not exposed but the Ardingly Sandstone forms low crags [4490 3975, 4415 3985] and river cliffs [4000 4435]. North of the Kent Water a low scarp is formed by the sandstone in the field [4320 4016] 700 yd (640 m) W. of Basing Farm. The bed dips 12° at 170° in the road [4296 4018] 200 yd (183 m) W. of that locality. North-westwards, however, the Ardingly Sandstone gives rise only to a low feature; exposures are rare and generally show flaggy sandstone. An old pit [4155 4145] ¼ mile (0·4 km) N.E. of Lady Cross Farm has now been partially filled in but a few feet of flaggy sandstone are still visible. This section, and a similar one in the stream bed [4178 4015], 300 yd (274 m) N.W. of Blockfield, lie very close to the junction with the Grinstead Clay.

North of the Kent Water, Ardingly Sandstone is again well exposed. Poor exposures of the lower member can be found in the lane banks [435 406 and 453 402] to the west and east of Scarletts. Crags of Ardingly Sandstone are well exposed in a field [4410 4035] near Furnace Farm and on the road bank alongside Furnace Pond. The sandstone dips 15° at 210° in the spillway [4362 4041] of the now dry Goudhurst Pond, 250 yd (229 m) N.W. of Basing Farm. C.R.B.

East Grinstead. In this area the Lower Tunbridge Wells Sand is about 110 ft (33·5 m) thick, the Ardingly Sandstone occupying the upper 50 to 60 ft (15·2–18·3 m). Boreholes at Dormans Park [397 405] and at Placelands, East Grinstead [393 384] proved 107 and 106 ft (32·6, 32·3 m) of Lower Tunbridge Wells Sand respectively.

Exposures in the lower part of the formation are rare. Silty sandstones and silts are poorly exposed in the railway cutting [400 380] 400 yd (366 m) E. of East Grinstead church and in a stream section [394 398] 900 yd (823 m) at 320° from Queen Victoria Hospital, East Grinstead.

By contrast the Ardingly Sandstone, which has provided the stone for most of the older buildings in East Grinstead, is very well exposed in numerous old quarries and in road cuttings. Old quarries at Hackenden [3942 3965 and 3973 3961], 750 yd (686 m) at 305° and 500 yd (457 m) at 327° from the above hospital, show deeply weathered iron-stained flaggy and thickly bedded sandstone for 6 ft (1·8 m), resting on 5 ft (1·5 m) of festoon-bedded greyish brown fine-grained sandstone resting on massive pale grey sandrock for 8 ft (2·4 m). Small exposures in the top of the Ardingly Sandstone are visible in some of the extensive pits in nearby Stonequarry Wood [399 394] where building stone was worked from beneath a capping of Grinstead Clay. The quarry

[3936 3879], 800 yd (731 m) at 234° from the Hospital, recorded by Topley (1875, p. 82) as providing the stone for St. Margaret's Convent (Orphanage) and which yielded a specimen of *Lepidotes mantelli* Agassiz (GSM 98624), is now completely degraded.

Small exposures of Ardingly Sandstone occur in road cuttings near The Larches [4039 3957]; 150 yd (137 m) N. of Hoskin's Farm [4026 3947]; 800 yd (732 m) W. of Gotwick Farm [4111 3935] and at Blackwell Hollow [3975 3859], 150 yd (137 m) W. of East Court. This last named, a sunken lane, shows the following composite section, in which plant debris, trace fossils and sedimentary structures are present throughout: deeply weathered flaggy sandstone passing down into finely festoon-bedded sandstone, 4½ ft (1·4 m); massive fine-grained grey sandstone with large scale trough cross-bedding, weathering to flaggy and thickly bedded (forms massive overhanging bed in cutting), 7 ft (2·1 m); silty, flaggy and thickly bedded sandstone with thin intercalations of grey silt, about 23 ft (7·0 m).

In the nearby railway cutting [396 383], 320 yd (293 m) N. of East Grinstead church, the following composite section was measured:

	Ft	in	(m)
Grinstead Clay:			
Grey shales weathering to clay	2	0	(0·60)
Finely interbedded siltstone and silty sandstone	0	6	(0·15)
Top Lower Tunbridge Wells Pebble Bed:			
Ripple-marked surface of coarse ferruginously stained sandstone (no pebbles seen) .. .. 0 to	0	0¼	(0·006)
Ardingly Sandstone:			
Silty sandstone with even, thick bedding	9	0	(2·74)
Very silty tough cream-coloured sandstone, exposed in part only .. .. about	7	0	(2·13)
Massive greyish brown and white sandrock .. about	30	0	(9·14)
Passing down into poorly exposed flaggy sandstones which in turn pass down into interbedded silts and sandstones of the lower part of the formation.			

The small faults recorded here by Topley (1875, p. 83, fig. 12) are no longer visible.

An old quarry [3983 3799] 200 yd (183 m) E. of the church exposes about 25 ft (7·6 m) of massive cross-bedded greyish white sandrock, and appears to have been worked at two levels. College Lane runs along the step between them and workings extending back from the lower face pass beneath the lane.

In a sunken lane [3966 3779] 180 yd (165 m) S. of the church are exposed 18 ft (5·5 m) of massive trough cross-bedded sandstone, with bedding units commonly 6 to 8 in (152–203 mm) thick. Vegetation and water seepage tend to obscure the sandrock nature of the lithology here. The same beds are poorly exposed in another sunken lane [3932 3765], 500 yd (457 m) S.W. of the church.

Charlwood Graben. Sandstones and siltstones of the lower part of the formation crop out in the drive at Charlwood [3933 3421] and in a lane [3973 3430] 420 yd (384 m) at 080° from Charlwood. Both sections are disturbed by faulting and the precise stratigraphical position within the formation is uncertain.

An old quarry [4040 3419] 120 yd (110 m) W. of Mudbrooks Farm shows the following section in Ardingly Sandstone: massive clean greyish white fine-grained sandstone, well jointed with cambering displacements, 5 ft (1·52 m); thinly bedded greyish brown fine-grained silty sandstone, 1½ ft (0·46 m); thin seams of highly carbonaceous, almost graphitic, dark grey clay with sandy partings and fresh mica plates, 0 to 2½ in (0–64 mm); massive clean greyish white fine-grained sandstone with lenses of small quartz pebbles (up to 1/10 in (3 mm)) in the upper part and small scattered pebbles throughout, 7 ft (2·13 m).

R.W.G.

Ashurstwood Outlier. The thicknesses of the members of the Lower Tunbridge Wells Sand at Ashurstwood are similar to those at East Grinstead.

Crags and low rocky outcrops of Ardingly Sandstone can be traced all round this outlier. The most spectacular are those that occur in the woods [414 371] to the south-east of Oakleigh. The crags are about 50 ft (15·2 m) high and have been accentuated by quarrying both here and at Brockhurst (Barrow 1915, p. 120). The Top Lower Tunbridge Wells Pebble Bed has been recorded in the old pit [4100 3758] alongside Shovelstrode Lane, 250 yd (229 m) S. of Worsted's Farm (Milner 1923b, p. 287, pl. 20b; Bazley and Bristow 1965, p. 315). Fragments of this pebble bed were found on the surface of the field [4330 3765] 700 yd (640 m) at 250° from Owlett's Farm.

Hartwell–Lyewood Common–Withyham–Eridge Green. Around Lyewood Common the Ardingly Sandstone is about 45 ft (13·7 m) thick out of a total of 130 ft (39·6 m) of Lower Tunbridge Wells Sand; at Eridge it is thicker and accounts for about a third of the 150 ft (45·7 m) of Lower Tunbridge Wells Sand.

The Hartwell Outlier was formerly shown as a fault-bounded block of Ashdown Beds, but except around St. Ives Farm the boundary is everywhere normal. A well [4712 3668] 200 yd (183 m) N.W. of Chartner's Farm penetrated 28 ft (8·5 m) of Lower Tunbridge Wells Sand before entering the Wadhurst Clay (Harvey and others 1964, 303/106). Thin flaggy sandstone can be seen in the old quarry and sunken lane immediately west of the farm. The dip in this outlier, about 2°E. into the Kent Water Valley, may be accounted for by cambering.

To the north of the Kent Water and Medway, the Lower Tunbridge Wells Sand outcrop is dissected by numerous tributaries. Massive laminated sandstone in the lower part of the succession is exposed in the river cliff [4927 3694] to the north of Withyham.

The Ardingly Sandstone makes a good feature, but only rarely gives rise to crags or rocky outcrops [4645 3855, 4675 3803, 4720 3825, 4767 3795, around 480 378 and 4825 3795, at 4848 3814 and 4877 3764]. Occasionally it is noted in stream sections and sunken lanes that the Ardingly Sandstone has lost its massive character and is flaggy [4758 3842, 4762 3831, 4848 3756, 4995 3690 and 5025 3696].

Milner (1923b, p. 288, pl. 20a) noted that the "glass sand" facies of the Lower Tunbridge Wells Sand (Ardingly Sandstone) was worked in a quarry [480 378] to the north of Hartfield.

South of the River Medway, as far as Mott's Mill, the Ardingly Sandstone is only poorly exposed, but can be traced by its feature. East of Mott's Mill the characteristic crags reappear.

Topley (1875, p. 89) described a section in the railway cutting [5090 3695] to the north of Stoneland's Farm. This section (totalling approximately 90 ft (27·4 m)), re-classified, is given below:

		Ft	in	(m)
Upper Tunbridge Wells Sand:				
Sandstone		4	0	(1·22)
Grinstead Clay:				
Clay and shale, red and green clay at the top	25 ft to	30	0	(7·62–
				9·14)
Lower Tunbridge Wells Sand:				
Sandstone		6	0	(1·83)
Green clay and shale		3	0	(0·91)
Sandstone		20	0	(6·10)
Shale		2	6	(0·76)
Sandstone		6	0	(1·83)
Clay and shale		3	0	(0·91)

Shaly sand	..	..	..	..	..	..	..	2	0	(0·61)
Hard sandstone		..	..	..	..	..	..	6	0	(1·83)
Clayey sand	..	..	..	..	..	..	..	2	0	(0·61)
Sandstone	..	..	..	..	..	..	..	10	0	(3·05)

Ribs of massive sandstone still appear in the lower part of the cutting and Grinstead Clay can be found at the top. Massive sandstone occurs at the roadside [5025 3593 and 5065 3624] and in an old quarry [5047 3605] south-west of Stoneland's Farm.

A section in the lower part of the Tunbridge Wells Sand, in the road cutting [4990 3584] east of Withyham, revealed thinly bedded flaggy sandstone dipping valleyward at 40°.

Massive sandstone was noted in a sunken lane [5032 3513] 300 yd (274 m) at 070° from Buckhurst Park, but farther east, at the cross-roads [514 350] south-west of Cherrygarden, and in the road bank [5164 3540] 200 yd (183 m) at 200° from Cherrygarden, horizontal, festoon-bedded sandstone is exposed. Similar festoon-bedded sandstone is found in the lane [5194 3534 and 5218 3554] to the south-east of the small fault which runs through Mott's Mill.

To the south of the stream (Mottsmill Stream) that flows north-eastwards through Mott's Mill are typical crags of Ardingly Sandstone. They outcrop in the grounds of Glen Andred and the railway cutting [530 359] where they dip 3°W. The crags at Leyswood have been attractively landscaped (Plate VIB). Rocks Wood [524 351] to the west of Leyswood, Penns in the Rocks and Jockey Wood [517 345 and 5165 3475] all provide good exposures of Ardingly Sandstone. The top of the Ardingly Sandstone is locally flaggy [5123 3646, 5228 3577 and 5225 3477].

To the east of the railway line the Ardingly Sandstone, capped with Grinstead Clay, is very well developed and is best known at Harrison's (see Topley 1875, p. 246, fig. 49) and Eridge Rocks [5325 3573 and 554 357] where the crags are especially impressive. The Harrison's Rock outcrop is used for climbing. Between these two localities the sandstone is indicated by a strong feature, coupled with low rocky protrusions [5477 3513]. The lower part of the Tunbridge Wells Sand, where exposed, is flaggy [542 348] or is a silty sandstone [5545 3535 and 5553 3543]. Locally the top of the Ardingly Sandstone is also flaggy [5405 3561 and 5408 3533].

Boarshead Outlier [540 330]. Formerly mapped as Ashdown Beds, it was first recognized by Herries (Abbott and Herries 1898, p. 451) that this outcrop might be Lower Tunbridge Wells Sand, although Abbott did not share his view.

The Lower Tunbridge Wells Sand overlies the Wadhurst Clay conformably and dips 5° at 010°. At The Rocks [537 328] and Bowle's Rocks [540 330] Ardingly Sandstone is well represented by rocky crags. The Bowle's Rocks outcrop, which is used as a training ground for climbers, is capped by a small outlier of Grinstead Clay. Casts of 'Cyrena' and Unio have been found in the sandstone here (Sweeting 1945, p. 154). Thin ferruginous flaggy sandstone can be found at the following localities: in the lane [5110 3298] at Summersales Farm; overlying massive sandstone which dips 20° at 020° in the old pit [5187 3315], 350 yd (320 m) N.E. of Gilridge Farm; and in the lane [5337 3314] ¼ mile (0·4 km) N.W. of Boarshead.

Bassett's Farm–Frienden Farm–Walter's Green Outlier and Blackham outliers. The first outlier has been almost dissected into two by a southward draining tributary to the Kent Water which flows past Bassett's Farm.

The Ardingly Sandstone is about 50 ft (15·2 m) thick and the total thickness of the Lower Tunbridge Wells Sand is about 120 ft (36·5 m).

Crags of Ardingly Sandstone occur all around the hill west of Bassett's Farm but are best developed on the north side (see Dines and others 1969). Lower Grinstead Clay and Cuckfield Stone cap the highest part of the hill. To the east of the farm

(A10306)

A.   WIDENED JOINTS IN ARDINGLY SANDSTONE AT LEYSWOOD
HOUSE NEAR ERIDGE STATION

PLATE VI

B.   ARDINGLY SANDSTONE ESCARPMENT AT BOWLES ROCKS
NEAR CROWBOROUGH

(A10322)

crags are not so well developed within this present district, but are more striking to the north (op. cit.). Massive sandstones can be seen in the old pit [498 412] 300 yd (274 m) at 110° from Bassett's Farm, where Grinstead Clay is present at the top but is not now exposed. A slipped block of sandstone was noted in the sunken lane [5019 4076] by Hobbs Hill. The sunken lane [510 408] west of Walter's Green affords another good exposure of the Ardingly Sandstone.

There are no exposures of the Lower Tunbridge Wells Sand in the three small outliers around Blackham. The valleyward dip of the easternmost one has already been mentioned (p. 64).

Ashurst–Langton Green–Tunbridge Wells–Pembury. It is within this outcrop that the Ardingly Sandstone (and also the Grinstead Clay) dies out as a lithological unit. It is not a progressive change from west to east but is fairly abrupt.

A well [5824 4021] sunk at Kelsey's Culverden Brewery in 1948 (manuscript record in the Institute of Geological Sciences) proved the following Tunbridge Wells Sand sequence to Wadhurst Clay and Ashdown Beds:

	Ft	in	(m)
Made Ground	4	0	(1·22)
Upper Tunbridge Wells Sand:			
Yellow rock	10	0	(3·05)
Grinstead Clay:			
Light grey hard clay	13	0	(3·96)
Yellow hard sandy clay	3	0	(0·91)
Dark brown clay	1	0	(0·30)
Lower Tunbridge Wells Sand:			
Yellow sandrock	10	0	(3·05)
Hard light grey sandrock	45	0	(13·72)
Hard dark grey clay	4	0	(1·22)
Hard light grey clay	13	0	(3·96)
White rock	5	0	(1·52)
Very hard layer	1	0	(0·30)
Less hard layer	5	0	(1·52)
Yellow sandrock	10	0	(3·05)
Light grey rock, very hard layers	12	0	(3·66)
Hard light grey rock	13	0	(3·96)
White rock	13	0	(3·96)
Wadhurst Clay	176	0	(53·65)
Ashdown Beds	19	0	(5·79)
Total depth	357	0	(108·81)

From the above record it is clear that the Ardingly Sandstone is there 55 ft (16·7 m) thick.

Around Ashurst there are few exposures in the Lower Tunbridge Wells Sand. The Ardingly Sandstone is not well developed and, except to the south-east, gives rise only to a poor feature. Roadside exposures near Stone Cross show massive [5245 3901] and flaggy [5259 3898 and 5262 3896] sandstone. An old quarry [522 382] to the south-west of New Park Wood shows massive sandstone weathering flaggy and dipping at 30° into the small valley which extends south-west from Stone Cross. Between this locality and Top Hill the Grinstead Clay is absent and it has not been possible to map accurately the Upper–Lower Tunbridge Wells Sand boundary. Eastwards, at Top Hill the Grinstead Clay reappears and can be traced to High Rocks. From Top Hill also the Ardingly Sandstone becomes more evident when traced eastwards. At first it forms only a feature but quarries [5400 3845, 5537 3837 and 5558 3847] and road sections [5460 3828] show local developments of sandrock facies. Elsewhere flaggy [5488 3867] or festoon-bedded [5455 3846] sandstones crop out. From High Rocks Farm [5590 3855] the sandrock crags can be traced to just south of Coldbath Farm, where they cross the valley and can be followed on the south side of the valley to High Rocks and from there to Ramslye Farm [5685 3795]. Just south of the farm the Tunbridge Wells Sand is thrown down against Ashdown Beds by the North Groombridge Fault. A section behind the farm revealed the junction of the Grinstead Clay and Ardingly Sandstone. No pebble bed was found here at the junction, nor was it located at High Rocks during archaeological excavations.

The Ardingly Sandstone when followed northwards around Fordcombe gives rise to a very poor feature. Sandrock was found in the roadside quarry [5205 4005] and as low rocky outcrops in Spring Hill Wood [5204 4033 and 5222 4037]. In a stream section [5314 4062] north-east of Fordcombe the beds immediately beneath the Grinstead Clay were flaggy sandstone. Similar beds at a slightly lower horizon were noted in a ditch [5370 4036] and a pit [5380 4066] to the west of Danemore Park. No pebble bed was found at the junction in a ditch section [5359 3988] 500 yd (457 m) W. of Shirley Hall.

Massive sandstone was evident in the numerous pits [5475 4003, 5445 3945, 5430 3935, 5408 3914, 5411 3909] around Langton Green and in road sections [5505 3957; around 5525 3990, 5627 4125, 5675 4100 and 5627 4028] to the north and west of Rusthall. Most of these exposures and those in the quarries [5557 4007 and 5605 4080] are in sandrock which is weathering to a flaggy sandstone.

The outcrop of Ardingly Sandstone around Rusthall, where it is overlain by Grinstead Clay, is well developed and gives rise to crags on the east side of the common, i.e. at Toad Rock [5691 3955] (see Topley 1875, p. 246, fig. 50; Gallois 1965b, pl. IIIA) and in the Happy Valley [5640 3922]. The rocky outcrops on Tunbridge Wells Common are notable for their intra-formational coarse pebbly beds up to 6 in (152 mm) thick. Outcrops of rock can be found at numerous localities north-westwards from the common: in the grounds of the hospital [5810 3997]; in the grounds [5805 4014, 5799 4025 and 5794 4034] to the west of the former Culverden Brewery (see p. 81); along Culverden Down Road [5758 4033 and 5752 4054]; at the Bennet Memorial Girls' School in pits dug through Grinstead Clay [5745 4060 and 5745 4050] and as rocky mounds in their grounds [573 407 and 5705 4013]; east of the school in the grounds [5784 4072] of The Down; and at the right-angled bend [5760 4096] in Reynolds Lane, where there is a good quarry section. No pebble bed was found at the junction of the Ardingly Sandstone and Grinstead Clay in the old pit [576 414] just north of the present district (Dines and others 1969, p. 47). To the east of the now demolished Culverden Brewery, sandrock outcrops were found in the cemetery [5845 4017]; in Shatter's Wood [5876 4087]; and as a continuous line of crags northward from a point [5844 4075] 1000 yd (914 m) at 020° from the hospital as far as the small fault at High Brooms on the edge of this district.

The railway cutting to the north-east of the hospital, in the lower part of the succession, still shows silts, flaggy sandstone and massive sandstone. The dip is variable. By the tunnel entrance [5859 4001] and the Grosvenor Bridge [5889 4046] the beds are horizontal. To the south-west of the bridge a dip of 25° at 310° was noted. Prestwich and Morris (1846, p. 397), who saw this section during the construction of the railway, recorded 45 ft (13·7 m) of strata which had a dip of 2°N.N.W; at one point there was a small reversed fault (op. cit., p. 402, fig. 4).

Eastwards from Tunbridge Wells both the Grinstead Clay and Ardingly Sandstone die out as mappable units. Rocky outcrops occur in gardens [5914 3971] and roadside exposures [5976 3980] to the east of Tunbridge Wells (both localities lie on either side of Sandrock Road). From here to Pembury the Ardingly Sandstone is traceable only by feature and in artificial exposures; a quarry [6105 4100]; a road cutting [6135

4130] opposite Pembury Hospital, and old quarries [6123 4145 and 6145 4160] to the north and west of the hospital on the adjacent Sevenoaks (287) Sheet (see Dines and others 1969, p. 45; Allen 1959, p. 298; 1962, pp. 241–3).

Three roadside exposures [6262 4087, 6298 4087 and 6302 4072] at Pembury of the sandrock are the most easterly occurrence of this facies that can be assigned to Ardingly Sandstone. To the south of these points a good feature marks its outcrop but eastwards it cannot be followed with certainty beyond Kipping's Cross.

The base of the Tunbridge Wells Sand is hidden under Head in Gregg's Wood [603 412] and east of Romford [648 412].

To the east of Kipping's Cross there are numerous roadside exposures [6507 4100, 6525 4048, 6632 4062, 6650 4058, 6706 4071, 6728 4067, 6713 4095 and 677 409] in silts, siltstone, sandstone and locally sandrock in the Tunbridge Wells Sand. A vertical bed of sandstone was noted in the stream bed [6621 4093] ½ mile (0·8 km) E. of Cork Pond. C.R.B.

Much of the higher ground in the vicinity of Lamberhurst Quarter is capped by Tunbridge Wells Sand. About 100 ft (30·5 m) of beds are preserved; fine white and ochreous sandstones and silts up to 50 ft (15·2 m) thick, at the base of the formation, are overlain by a thin seam of pale grey silty clay, which has been traced around the flanks of the unnamed hill north of Clay Hill Farmhouse, above which fine sandstones occur. An old quarry [6550 3315] 200 yd (183 m) at 345° from Clay Hill exposes the contact of the clay seam and underlying sandstones: topsoil, weathered rock 1 ft (0·3 m); grey, silty clay, seen up to 1 ft (0·3 m); fine, white cross-bedded sandstone, thinly bedded (3 to 6 in (76–152 mm)) with faint lamination and a 1-in (25-mm) ferruginous layer at top, seen up to 6 ft (1·8 m). A few feet of fine buff sandstones, nearer the base of the formation, are exposed in a disused quarry [6622 3864] 300 yd (274 m) at 310° of The Grange.

Several old workings occur in the outlier capping the hilltop east of Lindridge but all are now overgrown. The basal beds of the formation seen in a temporary exposure near Windmill Farm are described with the Wadhurst Clay (p. 69). E.R.S.-T.

The outcrop of Tunbridge Wells Sand lying between the Tunbridge Wells and North Groombridge faults is difficult to classify. In the west, around High Rocks, the succession is complete from Wadhurst Clay to Upper Tunbridge Wells Sand. In the east there is a normal Tunbridge Wells Sand/Wadhurst Clay junction. Between High Wood and Great Bayhall a well-marked feature, approximately 100 ft (30·5 m) above the base of the sands, probably represents the Ardingly Sandstone. Quarries [6067 3840, 6066 3854 and 6132 3884] and sunken lanes [615 394 and 6206 3954] expose massive sandrock along this feature, but nowhere are there crags comparable to those of High Rocks. The thin lenticular grey silty clay mapped at three localities on this outcrop could represent the feather edge of the Grinstead Clay. At two points [5998 3903 and 5988 3895] red clay was augered at the top of the clay.

Frant–Wadhurst–Lamberhurst. To the east of Frant, in the grounds of Ely Grange, it is possible to recognize the Ardingly Sandstone and Grinstead Clay. To the south and east of this locality some of the thin clay seams mapped, and some of the occurrences of massive sandstone, may be referable to these horizons (see below), but lithological correlation is not reliable as massive sandstones are known at both lower and higher horizons in the Tunbridge Wells Sand (see pp. 84, 97 and Dines and others 1969).

A borehole [5989 3615] in the grounds of Ely Grange started just below the base of the Ardingly Sandstone and proved 72 ft (21·9 m) of the lower part of the Lower Tunbridge Wells Sand before entering the Wadhurst Clay (manuscript record in the Institute of Geological Sciences, 303/130e).

Ardingly Sandstone forms low rocky outcrops on either side of the valley [5990 3615] to the east-north-east of Ely Grange and in the shaw [5996 3532] and stream

[6018 3508] to the west and south of Manor Farm. Old roadside excavations [5940 3563] show festoon-bedded massive sandstone. There are good exposures of Tunbridge Wells Sand in the cutting [5870 3588] of the Tunbridge Wells–Frant Road. To the east of Manor Farm the Ardingly Sandstone cannot readily be mapped. Massive sandstone may still be seen in the pit [6032 3534] just behind the farm, and flooring the lane which leads south-eastwards from here. Above this sandstone an impersistent silty clay, possibly the lateral equivalent of the Grinstead Clay, can be followed to Higham. A pit [6125 3576] in Higham Wood, situated below this clay seam, has thin flaggy sandstone overlying massive sandstone. Massive sandstone can be found at the entrance [6175 3567] to Higham. C.R.B.

A series of eight cored boreholes drilled in 1958 at Bartley Mill [631 656], 800 yd (732 m) at 196° from Wickhurst, for a reservoir site investigation, proved Lower Tunbridge Wells Sand resting on Wadhurst Clay. Detailed correlation between the boreholes was difficult owing to poor core recovery, but a complete sequence of Lower Tunbridge Wells Sand, about 140 ft (42·7 m), is probably present, the top 36 ft (11·0 m) consisting of massive sandstones which can be classified as Ardingly Sandstone. Details of two of the boreholes, Bartley Mill Nos. 1 and 2, which provide a full composite section, are given in Appendix I.

Below the Ardingly Sandstone the lower part of the formation consists predominantly of silts interbedded with silty sandstones and silty mudstones. The junction with the underlying Wadhurst Clay was not recovered in Borehole No. 2 but in the boreholes where it was recovered it is marked by a rapid upward transition from greenish grey, olive and red mottled mudstones to finely interbedded pale grey silts and silty mudstones. The Ardingly Sandstone, as elsewhere, consists of massive fine-grained quartzose sandstone containing a few fragmentary plant remains and scattered small pebbles, usually of quartz, up to $\frac{1}{4}$ in (6 mm) in size. R.W.G.

It is not known with certainty to what horizon other quarry exposures [6170 3622 and 6285 3595] of massive sandstone belong, but they appear to be high (over 250 ft (76·2 m) from the base) in the succession. C.R.B.

A discontinuous silty clay can be traced around Buss's Green and Wickhurst. A line of seepages in the field [632 353] to the west of Wickhurst may indicate the presence of a lower clay seam. A somewhat thicker seam is found in Brookland Wood [616 351], $\frac{1}{2}$ mile (0·8 km) E. of Barelands, where the clay has been worked in pits. A red top to the clay was proved at two points [6166 3510 and 6176 3506]. Its relationship to the clay seam on the opposite side of the valley is uncertain, but it appears to be at a lower horizon than this unless there is faulting between the two outcrops. C.R.B.

Between the North and South Groombridge faults, west of Lamberhurst, up to 150 ft (45·7 m) of the formation are preserved, consisting largely of fine white and ochreous sandstones with smaller proportions of silt and silty clay. A small roadside section [6694 3668] 950 yd (869 m) at 257° from Mount Pleasant exposed up to 4 ft (1·2 m) of pale, medium-grained sandstone, showing foresets inclined to the north-north-west. South of the Groombridge faults lies a wide belt of Tunbridge Wells Sand country in which the succession is very similar to that in the outcrops north of Lamberhurst. The lower 50 to 80 ft (15·2–24·4 m) of the formation comprise fine, white and buff sandstones with subordinate grey silts; sandstones at the top of this group usually give rise to a steep feature in the valley sides hereabouts, the top of which marks the outcrop of the overlying bed of pale grey, often red mottled, silty clay. The clay, which is up to 20 ft (6·1 m) thick, is persistent over much of this district, its outcrop being frequently repeated by faulting. Fine buff and ferruginous sandstones with silts above the clay seam make up a total of about 150 ft (45·7 m) of the formation preserved.

An old quarry [6597 3644] 360 yd (329 m) at 330° of Furnace Mill exposes up to 6 ft (1·8 m) of thinly bedded, pale grey, very fine-grained and silty sandstones, near the base of the formation. A small roadside pit [6771 3566] 350 yd (320 m) at 350° of

Spray Hill is much degraded but shows up to 6 ft (1·8 m) of ochreous grey silt with thin ferruginous sandstones above. Extensive old quarries at Lamberhurst Down [6745 3543] were worked through the clay seam exposing the sandstones beneath; the workings are now much overgrown and degraded but small sections in sandstones can still be seen.

South of the Chingley Fault, a well at Bewkes [6567 3473], 350 yd (320 m) at 295° of Crowhurst, proved only 25 ft (7·6 m) of soil and Tunbridge Wells Sand above Wadhurst Clay in the valley bottom.

North of the Hook River between Cousley Wood and Lower Cousley Wood the clay seam is somewhat thinner, but the sandstone feature beneath is usually well developed. South of Langham [6565 3369], where the outcrops are repeated by closely spaced faults, a stepped or terraced topography results. Massive buff sandstones above the clay seam are exposed up to 10 ft (3·05 m) in the sunken lane [6690 3376] 500 yd (457 m) at 100° of Ladymeads.

Several of the trial boreholes sunk at the proposed reservoir site in the Bewl Valley penetrated the lower part of the formation, which here comprises alternations of fine sandstones and silts with thin silty clay. The detailed lithology varies greatly from borehole to borehole and it is difficult to correlate between them. The log of one borehole is given below.

Ticehurst Reservoir No. 17, [6733 3398] 1961. O.D. 332 ft (101·2 m)

	Thickness			Depth		
	Ft	in	(m)	Ft	in	(m)
Tunbridge Wells Sand:						
No core taken	13	6	(4·11)	13	6	(4·11)
Pale grey silt	1	0	(0·30)	14	6	(4·42)
Fine-grained, cross-bedded sandstone ..	1	5	(0·43)	15	11	(4·85)
Grey and ochreous silt	0	3	(0·08)	16	2	(4·93)
Grey and ochreous, fine silty sandstone ..	3	6	(1·07)	19	8	(5·99)
Core lost	0	4	(0·10)	20	0	(6·10)
Friable, ochreous, grey and white, fine, cross-bedded sandstone with a few thin silty partings with plant debris	9	0	(2·74)	29	0	(8·84)
Core lost	1	0	(0·30)	30	0	(9·14)
Friable, grey and white, fine sandstone with silty partings	7	3	(2·21)	37	3	(11·35)
Dark grey silt with thin cross-bedded lenticles of very fine white sand	1	9	(0·53)	39	0	(11·89)
Core lost	0	6	(0·15)	39	6	(12·04)
Dark grey silt, shaly and sandy in part ..	2	10	(0·86)	42	4	(12·90)
Compact fine grey sandstone	0	9	(0·23)	43	1	(13·13)
Violet-grey silt, clayey at base	1	7	(0·48)	44	8	(13·61)
Compact, but porous, fine ochreous sandstone..	4	2	(1·27)	48	10	(14·88)
Laminated grey sandy silt	0	3	(0·08)	49	1	(14·96)
Fine, grey and ochreous, cross-bedded sandstone	3	1	(0·94)	52	2	(15·90)
Grey silt with some very fine sand	0	10	(0·25)	53	0	(16·15)
Fine, white and ochreous, cross-bedded silty sandstone penetrated by fine vertical pores (? rootlets)	4	4	(1·32)	57	4	(17·48)
Core lost	0	8	(0·20)	58	0	(17·68)
Grey sandy silt	1	0	(0·30)	59	0	(17·98)
Core lost	1	0	(0·30)	60	0	(18·29)
Compact, dark grey sandy silt, red-stained with fragments of carbonaceous plant debris ..	3	0	(0·91)	63	0	(19·20)

	Thickness			Depth		
	Ft	in	(m)	Ft	in	(m)
Pale to dark grey silt, clayey at top	4	2	(1·27)	67	2	(20·47)
Compact, dark greenish grey, red-stained silty clay	2	0	(0·61)	69	2	(21·08)
Compact, violet-grey silt with plant debris ..	3	0	(0·91)	72	2	(21·99)
Dark grey and greenish grey clayey silt, red-stained in part, some fine sand, micaceous ..	4	1	(1·24)	76	3	(23·24)
Fine, white, cross-bedded sandstone and grey silt	2	4	(0·71)	78	7	(23·95)
Compact dark violet-grey silt	2	9	(0·84)	81	4	(24·79)
Fine, grey and white, cross-bedded sandstone ..	2	4	(0·71)	83	8	(25·50)
Dark grey cross-bedded silt with sand laminae passing down into greenish grey silt ..	1	4	(0·41)	85	0	(25·91)
Pale grey, fine sandy silt, ½-in (13-mm) ferruginous band at base	1	5	(0·43)	86	5	(26·34)
Fine, ochreous, cross-bedded sandstone ..	3	11	(1·19)	90	4	(27·53)
Ochreous, khaki silty clay..	0	8	(0·20)	91	0	(27·74)
Fine, grey and white, silty sandstone	7	4	(2·24)	98	4	(29·97)
Grey silt	2	6	(0·76)	100	10	(30·73)
Fine, white and grey, silty sandstone, iron stained	4	0	(1·22)	104	10	(31·95)
Fine grey sandstone with fragments of carbonaceous plant debris	0	4	(0·10)	105	2	(32·05)
Greyish green silty clay	0	6	(0·15)	105	8	(32·21)
Core lost	0	4	(0·10)	106	0	(32·31)
Medium, white, cross-bedded sandstone, porous	2	0	(0·61)	108	0	(32·92)
Ochreous grey silty clay	0	6	(0·15)	108	6	(33·07)
Ochreous sandstone	0	2	(0·05)	108	8	(33·12)
Iron-stained grey silt with fine sand	3	8	(1·12)	112	4	(34·24)
Compact fine grey ferruginous silty sandstone ..	5	10	(1·78)	118	2	(36·02)
Compact greenish grey silt with some fine sand ..	4	2	(1·27)	122	4	(36·29)
Fine grey silty sandstone	0	10	(0·25)	123	2	(37·54)
Olive-green clayey silt	1	10	(0·56)	125	0	(38·1)
Fine, grey, cross-bedded silty sandstone.. ..	5	7	(1·70)	130	7	(39·8)

?Wadhurst Clay:

Dark greyish green finely laminated silty clay ..	1	4	(0·41)	131	11	(40·21)
Laminated grey silt with fine sand	1	9	(0·53)	133	8	(40·74)
Olive-green silty clay	0	2	(0·05)	133	10	(40·79)
Core lost	1	2	(0·36)	135	0	(41·15)

A lane-side exposure [6542 3313], 600 yd (549 m) at 280° of Little Butts, shows up to 5 ft (1·5 m) of fine, ochreous cross-bedded sandstone inclined 5° at 020°. E.R.S.-T.

In Eridge Old Park, 12-ft (3·6-m) crags of massive quartzose sandstone are exposed in an old quarry [5763 3432] and on the hillside to the south [5753 3406]; the sandstone dips at about 3°S. and a feature about 20 ft (6·1 m) high can be clearly traced. Not far to the east [5815 3429] 20-ft (6·1-m) crags of similar sandstone are exposed. This massive sandstone is about 60 ft (18·3 m) above the base of the formation and may be equivalent to the Ardingly Sandstone. No clay that might be the equivalent of the Grinstead Clay could be found at the top. The feature of a massive sandstone at a similar horizon can be traced around Shernfold Park [5907 3491]. In Great Wood [5964 3396] 18 ft (5·5 m) of a massive well-jointed sandstone, probably equivalent to the above, are exposed in an old quarry; and crags of similar size are present along the striking feature that can be traced as far as Lightlands [5965 3335].

In Rocks Wood 15-ft (4·6-m) crags of massive sandstone crop out 600 yd (549 m) at 343° from Saxonbury Hill Tower [5762 3350]. Above this sandstone a thin silty clay was traced for some distance and could possibly be the equivalent to the Grinstead Clay (see p. 43 and Reeves 1948, p. 259). The clay is a maximum of 10 ft (3·1 m) thick and at an horizon about 80 ft (24·4 m) above the base of the formation. At the side of the A267 road in Saxonbury Wood is exposed an 8-ft (2·4-m) crag of massive sandstone of the same horizon as those in Rocks Wood [5812 3311]. In Nap Wood [5880 3280] paths cut through the lower 60 ft (18·3 m) of the Tunbridge Wells Sand indicate that silts are most common, with only thin sandstone beds. The silts form poorly draining ground, where reeds flourish.

Between Earlye Farm [5975 3293] and Buss's Green [6450 3490] three impersistent clay seams, commonly slightly silty, are present. These are at horizons about 40, 80 and 135 ft (12·2, 24·4, 41·1 m) above the base of the formation. All are less than 10 ft (3·0 m) thick, and the highest is the most persistent. All three clays were traced in Camden Wood [6200 3460] and around Buckhurst Manor [6100 3257]. The presence of many small faults in this region makes the tracing of these clays difficult, but generally the clay with a base about 80 ft (24·4 m) above the Wadhurst Clay has beneath it a more massive sandstone than is common elsewhere in the formation in this area. This thin silty clay may be equivalent to the Grinstead Clay, and the massive sandstone to the Ardingly Sandstone.

Massive well-jointed sandstone crops out beneath the middle clay 100 yd (91 m) at 100° from Earlye Farm [5985 3290]. Some 270 yd (247 m) at 042° from Riverhall, in a stream section [6071 3351], 4 ft (1·2 m) of massive yellow sandstone are exposed above 3 ft (0·9 m) of yellow silts and fine-grained sands. About 100 yd (91 m) away [6079 3357] a quarry section shows 20 ft (6·1 m) of massive cross-bedded sandstone with thin silt layers. Some sandstone surfaces show ripple marking and in the east side of the quarry there is a 4-ft (1·2-m) deep channel cut into the massive sandstone and filled with medium grey silt which contains lenses of sandstone. The massive sandstone thins rapidly towards the west. Plant fragments are present all through the sequence.

In the stream 600 yd (549 m) at 314° from The Mount there are many small sections of light grey silts and flaggy sandstones which are affected by valley bulging.

In the railway line opposite the Rockrobin [6223 3300] there are several good sections in massive sandstone up to 10 ft (3·0 m) thick. This sandstone is probably about 80 ft (24·4 m) above the base of the formation.

An old quarry at Four Acre Wood [6260 3327], 450 yd (411 m) at 050° from the Rockrobin, exposes about 30 ft (9·1 m) of massive light grey cross-bedded sandstones. In places these are strongly ripple marked and three dominant sets of joint planes are well developed trending approximately N. to S., N.N.W. to S.S.E. and N.N.E. to S.S.W. The crest of a minor anticlinal structure with an axis trending east to west is exposed in this quarry, which is now disused and largely overgrown; a section was photographed previously (Milner 1924, p. 390). This sandstone is faulted on all sides. It is probably about 60 ft (18·3 m) above the base of the formation. A thin clay seam above it, once worked 800 yd (731 m) E. in Pook Pit, is possibly equivalent to the Grinstead Clay.   R.A.B.

South of the Hook River, in a much-faulted block of country, about 80 ft (24·4 m) of fine white and ochreous sandstones at the base of the formation are overlain by a fairly persistent clay seam of variable thickness up to 20 ft (6·1 m); the higher beds present are fine sandstones with grey silt. An old quarry [6723 3205] 250 yd (229 m) at 020° of Claphatch shows the following descending sequence in the lower part of the formation: buff sandstone and grey silt (obscured), 3 ft (0·9 m); fine, ochreous sandstone, 1 ft 9 in (0·5 m); fine, ochreous cross-bedded silty sandstone thinly bedded in layers ½ to 1 in (13–25 mm) thick, 2 ft (0·6 m); massive fine white and ochreous sandstone up to 4 ft (1·2 m).

Several large old pits, sited on the clay seam, occur hereabouts near Foxhole [6530 3175], Little Whiligh [6575 3165] and Birchett's Green [6675 3150]; it is probable that the clay extracted was used for 'marling'.   E.R.S.-T.

Danehill–Sheffield Park–Maresfield–Buxted–Hadlow Down. At Freshfield Lane Brickworks [382 264] on the adjacent Horsham (302) Sheet the base of the Cuckfield Stone channels down to cut out all but a few feet of the underlying Lower Grinstead Clay (Gallois 1964, p. 362). Evidence farther south on the Horsham (302) Sheet suggests that the base of the Cuckfield Stone progressively cuts down into the underlying beds as it is traced eastwards. East of a line running from Danehill to Sheffield Park the Cuckfield Stone rests directly on Ardingly Sandstone and, in the absence of the intervening Lower Grinstead Clay, the lithological differences between these two sandstones are insufficient for them to be separated. West of Danehill [at 394 275] and around Warr's Farm, Chailey [395 230], the outcrops shown as Cuckfield Stone resting on Ardingly Sandstone are in continuity with outcrops on the Horsham (302) Sheet where the Lower Grinstead Clay is still present. In the three inliers S.W. and S. of Danehurst [402 266] Grinstead Clay rests on massive sandstones of 'sandrock' lithology. These inliers are separated by faulting from the outcrops at Danehill and Chailey and there is no direct evidence of the precise stratigraphical position of the sandstones. If they are interpreted as Ardingly Sandstone then it is necessary to explain how the Cuckfield Stone can die out (from 20 ft (6·1 m) thick at Freshfield Lane Brickworks) and how the total Grinstead Clay thickness can be reduced to 20 to 25 ft (6·1–7·6 m) (from 44 ft (13·4 m) at the brickworks) in the distance of 1100 yd (1 km) which separates the brickworks from the nearest of the three inliers. If the sandstones are interpreted as Cuckfield Stone then it is necessary to explain how the distinctive festoon bedding and dark brown staining (possibly indicating decalcification) of the Cuckfield Stone seen at the brickworks, at Danehill and at Chailey, could be replaced in the inliers by massive greyish white sandstone.

In the three inliers and east of the Danehill to Sheffield Park line these sandstones have been coloured on the map as Ardingly Sandstone and the possibility that they may include beds equivalent to the Cuckfield Stone of adjacent areas has been left open.

Ardingly Sandstone *sensu stricto*, consisting of 7 ft (2·1 m) of massive fine-grained white sandstone, is exposed in an old quarry [3922 2711] 650 yd (594 m) at 232° from Danehill church.

In the inlier [394 260], 1100 yd (1 km) S.W. of Danehurst, a few poor exposures of flaggy sandstone occur in the stream course. A prominent feature immediately below the junction with the Grinstead Clay, about 20 ft (6·1 m) high and yielding much flaggy sandstone debris, may be formed by Cuckfield Stone resting on Ardingly Sandstone. Sandstone has been dug for building purposes from beneath a cover of Grinstead Clay in old pits [3948 2627 and 3965 2638] but no exposures are now present. Fragments of coarse ferruginous sandstone occur adjacent to the junction with the Grinstead Clay at several localities and these may be the equivalent of either the Top Lower Tunbridge Wells Pebble Bed or the Cuckfield Pebble Bed.

In the adjacent inlier [401 257] massive greyish white sandstones crop out along both sides of the valley. A small gorge cut along a major joint [4001 2568], 650 yd (594 m) at 265° from Slider's Farm, exposes 18 ft (5·5 m) of massive sandstone overlain by cambered clays. A similar section showing 16 ft (4·9 m) of sandstone occurs 80 yd (73 m) N.E. [4006 2571] in another stream course.

On the opposite side of the valley [4030 2578], 250 yd (229 m) at 260° from Slider's Farm, natural outcrops of sandrock form crags up to 8 ft (2·4 m) high.

Numerous pits have been dug through the Grinstead Clay on both sides of the valley to work the sandstone for building purposes. One such pit [3979 2527], 720 yd (658 m) at 330° from Ketche's Farm, still shows massive greyish white sandstone overlain by yellowish brown clays. Cambering of the clays has obscured the junction but debris in the adjacent fields suggests that it is marked by a thin bed of coarse ferruginous sandstone.   R.W.G.

In the outcrops to the east and south-east of Sheffield Green the thickness of the Lower Tunbridge Wells Sand is about 150 ft (45·7 m).

Massive Ardingly Sandstone floors the stream valley [407 253] to the west of Sheffield Green. North-east of the green there is a wider tract with rocky outcrops: on Brookers Rough [around 4130 2623 and 414 261]; in Beechy Wood [4131 2577]; and in Sheffield Forest [around 418 261 and 4200 2615]. Roadside exposures [4289 2614] of this sandstone to the north of Searles are inclined 2° at 170°. An old pit [4335 2600] 150 yd (137 m) N.E. of Searles exposes sandrock weathering flaggy. In a field southeast of Searles there is a low rocky outcrop [4344 2555]; massive sandstone, dipping 2° at 170°, floors the stream [4367 2535 and 4367 2545] draining southwards from Searles Lake. The Grinstead Clay is underlain by thin flaggy siltstone in the roadbank [4307 2536] to the south of Searles.

Around Courtlands a faulted outlier of Ardingly Sandstone is well exposed, both in the lane leading past the farm and in the fields to the east [4468 2638 and around 4472 2644]. The dip of the sandstone in the lane is 2° at 240°. South of Old Forge Farm are numerous small exposures and rocky crags of Ardingly Sandstone: around Woodcock Farm [444 261] where the beds are inclined 4° at 190°; around Black Ven [445 256] and to the south and west of Horney Common [448 258, 4487 2566, 4515 2565, 4512 2544, 4535 2535, 4545 2515, 4550 2493 and around 4505 2487]. In the vicinity of the junction of the A22 and B2026 roads the sandstone is again well exposed: south-west of the junction, in a cave [4608 2515] excavated in the crags; west of the B2026 around Lampool [462 256]; east of the B2026 in the grounds of the Doma Farm Nursery [464 255]; north-west of Hendall in Rock Wood [470 259]; and in Furnace Wood [4725 2607 and 4725 2588].

To the south-east of the Fletching Fault there is a wide outcrop of Ardingly Sandstone with numerous exposures to the south and west of Maresfield: around Flitterbanks [4600 2475]; in the sunken lane [4560 2453] to the south-west of Flitterbanks; in the field [4492 2405] and in Forge Wood [4517 2363] east and south-east of Down Street; in Maresfield Park [4594 2428 and 4608 2443]; in the centre of Maresfield [4684 2405]; on the southern bank of the lake [4630 2295] and on Black Down [4650 2297] west of Budlett's Common; in the grounds of the Warren [476 233]; in Budlett Wood [469 228, 4666 2275 and 4675 2260]; and in Longwood Gill [4655 2220]. To the south of these localities the Ardingly Sandstone forms a well-known line of crags on the adjacent Lewes (319) Sheet around Uckfield.

The triangular fault-bounded outcrop of Lower Tunbridge Wells Sands to the east of Maresfield has only the lower part of the succession present. The regional dip is about 1½° at 210°. Some 42 ft (12·8 m) of sands and silts were proved resting on Wadhurst Clay in a well at Gate House [4720 2404]. Flaggy sandstone in the road bank [4716 2387] just south of Gate House is inclined 2° N.W. Two trench sections [4779 2387 and 4796 2436] at Five Ashes revealed 10 to 15 ft (3·1–4·6 m) of ferruginous sands, sandstone and siltstones. Roadside exposures [4893 2408] in thin flaggy sandstone to the south of New House dip 1° at 200°. Thicker bedded sandstone weathering flaggy is present in the road bank [488 254] opposite Hogg House.

There are several small faulted outliers between the Hendall Wood Fault and Heron's Ghyll. The outlier immediately south of Heron's Ghyll has the most complete succession, with a total thickness of about 125 ft (38·1 m) of the Lower Tunbridge Wells Sand, of which 40 ft (12·2 m) are Ardingly Sandstone. The dip is about 8° at 160°. Crags of sandrock occur all along the outcrop and are capped by Grinstead Clay.

To the south of the above locality is the Stonehouse Outlier, which has a general inclination of about 4°S. The only exposure, of thin sandstones and silts, is to be found in the lane [488 255] to the north-west of Stonehouse.

Between the Greenhurst and Hendall Wood faults the Lower Tunbridge Wells Sand outcrop is complicated by splinter faults. Crags and rocky outcrops of Ardingly Sandstone are common: at the roadside [485 251] 400 yd (366 m) N. of Bevingford;

near Tudor Rocks [493 251]; and around Hermitage Farm [495 240] (see Topley 1875, p. 247, fig. 52).

At Buxted, Tunbridge Wells Sand in the railway cutting overlies Wadhurst Clay conformably, and no Ashdown Beds as shown in fig. 7 accompanying Gould's notes (Topley 1875, p. 66) could be proved therein.   C.R.B.

The Buxted to Hadlow Down Outlier, faulted along most of its southern and northern boundaries, has many small sections in the roadsides and streams which show sandstones and silts that generally have weathered light yellowish brown in colour. A total of about 190 ft (57·9 m) of the formation is exposed and the alternation of massive sandstones and silts of which it is comprised is admirably indicated by the features all over the area. Springs are common at the top of the silt or silty clay horizons within the sequence.

Some 300 yd (274 m) at 263° from Greenhurst in an old quarry are 12 ft (3·7 m) of cross-bedded massive sandstone of the Ardingly Sandstone. The top surface shows signs of channelling but there is no pebble bed. Above is much weathered yellowish grey silty clay, the Grinstead Clay.

At 50 yd (46 m) S. from Popeswood Farm [5115 2330] excavations for a new water reservoir showed the following downward succession: massive sandstone grading downwards into light grey silt, with nodules of clay ironstone up to 2 in (51 mm) thick, about 3 ft (0·9 m); massive light grey sand, 2 ft 3 in (0·7 m); light grey silt, 2 ft 3 in (0·7 m); light grey sand, ripple marked, 2 ft (0·6 m). Streaks of black carbonaceous material were common throughout. Though massive and generally light grey when fresh, the sandstones become flaggy and light yellowish brown on weathering.

Two thin clay seams, about 5 ft (1·5 m) thick, were traced around Hadlow Down [5320 2400]. The base of the lower and most persistent of these is probably about 145 ft (44·2 m) and the upper about 170 ft (51·8 m) above the Wadhurst Clay.

Roadworks in 1964 on the A267 to the south of Five Ashes near Hodges [5550 2410] revealed the junction between a clay and a sandstone. The descending sequence is: medium grey clay, very weathered, 1 ft (0·3 m); slightly silty clay, 2½ in (63 mm); clay ironstone, variable thickness, 1½ in (38 mm); light grey clay, 2 in (51 mm); yellow sandstone, base coarse grained, 1 in (25 mm); light grey silty clay, 3 in (76 mm); massive yellow sandstones with thin silty partings, 3 ft (0·9 m). The above clay may be the equivalent of the Grinstead Clay. The sandstone below becomes quite massive; and 400 yd (366 m) S. from Hodges 8 ft (2·4 m) of massive quartzose silty sandstones were exposed in the road section and showed a shallow anticlinal structure [5538 2379] with an axis trending approximately east to west.

Another clay, mottled red and yellow, is present 300 yd (274 m) S. from Allen's Farm. This is probably about 80 ft (24·4 m) above the base of the formation.

Rotherfield–Wadhurst–Stonegate area. Several small outliers are present in this area as thin cappings to hills. Most comprise less than 50 ft (15·2 m) of Tunbridge Wells Sand. Rotherhurst [5540 2850] is built on the edge of Cottage Hill, which is capped by about 70 ft (21·3 m) of Tunbridge Wells Sand. The road here [5538 3870] is cut through about 12 ft (3·7 m) of massive sandstones and thin silts and in the fields below Rotherhurst the well-developed spring line is typical of the Tunbridge Wells Sand and Wadhurst Clay junction.

Road widening of the A267 at Green Hill [5755 2934] in 1965 made temporary sections in the silts and fine-grained sandstones that are typical of the lower 30 ft (9·1 m) of the formation; red clays of the Wadhurst Clay were exposed below.

On Bestbeech Hill massive sandstones are exposed in several places, and all dip at about 10° S.E.; they were once quarried, some 12 ft (3·7 m) of massive sandstones still showing [6189 3150]. In the road section opposite the Best Beech Hotel there are flaggy fine-grained sandstones.

(A 10317)

A. Massive top Ashdown Beds in Under Rockes Wood, Butcher's Cross, near Mayfield

PLATE VII

B. Massive Ardingly Sandstone at Eridge Rocks

(A 10303)

Mayfield area. The western part of this outlier is much faulted, and the precise thickness of Tunbridge Wells Sand is uncertain. At the western end the formation probably reaches 100 ft (30·5 m), the greatest thickness in the inlier. Massive sandstones are exposed in the road [5625 2672] 200 yd (183 m) at 003° from Horleigh Green Farm. Along the boundary of Wet Wood 400 yd (366 m) E. [5660 2665] a spring line is well developed and indicates a possible fault. At Woolbridge, a bed of clay within the sequence is exposed in the banks of the River Rother [5674 2688]. It is mottled red, brown and grey, not more than 10 ft (3·0 m) thick, and slightly silty in part; the horizon is probably only about 20 ft (6·1 m) above the base of the formation.

The town of Mayfield is built on massive sandstone of the Tunbridge Wells Sand. This is possibly equivalent to the Ardingly Sandstone, but there is no development of the Grinstead Clay. Crags of a massive sandstone, only about 30 ft (9·1 m) above the base of the formation, are exposed along most of the roads, and in a quarry 450 yd (411 m) at 247° from the Congregational Church about 20 ft (6·1 m) of such light grey sandstone, weathering to light brown, contain plant debris.

Some 300 yd (274 m) at 270° from Little Bainden 12 ft (3·7 m) of massive sandstones are seen along the road [6025 2633]. About 100 yd (91 m) N. of Gillhope a section in an old saw pit [6110 2620] shows in descending sequence: light yellowish brown sandy silts, 2 ft (0·6 m); coarse-grained massive light yellowish brown sandstone with a 1-in (25-mm) silt parting, 6 ft (1·8 m); light yellowish brown cross-bedded sandstone, upper part flaggy, becoming more massive downwards, 10 ft (3·0 m). Finely disseminated plant remains are common in all the sandstones. This section is again only about 30 ft (9·1 m) above the Wadhurst Clay. Compared with other areas the Tunbridge Wells Sand is hereabouts generally more massive near its base.

In Hawkesden Park Wood an impersistent seam of red clay, less than 10 ft (3·0 m) thick, was traced for a short distance within the Tunbridge Wells Sand [6194 2630]; it lies about 75 ft (22·9 m) above the base of the formation.

Five Ashes, Burwash area. A newly dug hole for a cess pit [6188 2419] 700 yd (640 m) at 317° from Paine's Cross revealed 9 ft (2·74 m) of light grey silt with many thin clay seams on 1 ft (0·3 m) of stiff light grey clay; at the surface this rock is relatively impermeable to water and causes marshy ground along its outcrop.

Around Burwash there are several outliers of Tunbridge Wells Sand, most being less than 50 ft (15·2 m) in total thickness. The outlier at Holmshurst [6435 2545] is about 105 ft (32·0 m) thick; fine-grained yellow sands and silts are exposed along the lanes. A massive sandstone feature forms the hill above the Kicking Donkey Inn [6470 2595] and the bed may be the equivalent, in this area, of the Ardingly Sandstone.   R.A.B.

GRINSTEAD CLAY

(including Cuckfield Stone)

Blockfield Wood–Kent Water Valley. In Blockfield Wood the Grinstead Clay is about 50 ft (15·2 m) thick, with no development of the Cuckfield Stone. This outcrop was formerly thought to be Wadhurst Clay (see p. 77). Numerous pits, many flooded, pock the Grinstead Clay outcrop, but there are now no exposures. Red clay can be followed continuously beneath the base of the Upper Tunbridge Wells Sand.

There are two small faulted outliers around Basing Farm. A temporary section [4367 4032] for some farm buildings in 1963 revealed the junction of the Upper Tunbridge Wells Sand and Grinstead Clay. The top of the clay was not reddened.

There are no exposures in the outliers at Furness Farm or in Reading Wood [444 398], although several pits have been opened in this clay in the past.   C.R.B.

East Grinstead. The Grinstead Clay reaches a maximum of about 50 to 60 ft (15·2–18·3 m) in thickness in this area, but only the lower part of the formation is exposed on the present map. A borehole at Dormans Park [397 405] proved 82 ft (25 m) of Grinstead Clay, but this thickness probably includes some Upper Tunbridge Wells Sand. Cuckfield Stone is absent in this area.

Poor exposures of deeply weathered clay occur in some of the pits in Stonequarry Wood [399 394] (see p. 77) and in the pit near the orphanage [3936 3879] (see p. 78). One of the Stonequarry Wood pits [3990 3949], 300 yd (274 m) N. of Queen Victoria Hospital shows: Grinstead Clay, comprising stiff yellow and grey clay, 4 ft (1·2 m); Top Lower Tunbridge Wells Pebble Bed, with a megaripple surface on medium-grained sandstone, coarse on ripple crests, no pebbles seen, 1½ in (38 mm); Ardingly Sandstone, thickly bedded, fine-grained and iron-stained, 6 in (152 mm).

The old brickworks at Hackenden [3945 3932], 550 yd (503 m) at 282° from the Hospital, has now been built over. A section exposed in the back of a small landslip [3940 3938] on the railway line immediately north of Hackenden Brickworks showed disturbed grey shales with a thin bed of 'Cyrena' limestone, a thin bed of limestone composed almost entirely of ostracods and a bed of iron-stained silty mudstone, up to 1 in (25 mm) thick, packed with *Equisetites* stems, rhizomes and rootlets, many in position of growth. This soil bed, for which the name Hackenden Soil Bed is proposed, occurs about 20 to 25 ft (6·1–7·6 m) above the base of the Grinstead Clay at a level approximately the same as that of the Cuckfield Stone of the Ashurstwood and West Hoathly areas. The rhizomes of the *Equisetites* here are much smaller than the *E. lyelli* (Mantell) of the High Brooms Soil Bed (see p. 42), being up to 0·3 in (8 mm) diameter (commonly 0·4 by 0·2 in (10 by 5 mm) when crushed) compared to up to 0·5 in (13 mm) diameter (0·7 by 0·4 in crushed (18 by 10 mm)) for examples of the latter bed collected at Balcombe, Cuckfield and Freshfield Lane, Danehill.   R.W.G.

Ashurstwood. There are four main outliers, three of which are capped by Cuckfield Stone. There is no Upper Grinstead Clay present. The sandstones are regarded as Cuckfield Stone rather than Upper Tunbridge Wells Sand as they lie on the line of the Frienden Farm–Lyewood Common–Horsted Keynes outcrops, where the Cuckfield Stone is well developed. The thickness of the Grinstead Clay, 20 to 30 ft (6·1–9·1 m), is comparable to that of the Lower Grinstead Clay of the above localities.

The junction of the Grinstead Clay and Ardingly Sandstone in the pit at Shovel-strode Lane is described above (p. 79).

A ditch section [4220 3718] in the Cuckfield Stone at Ashurstwood revealed thin flaggy siltstone. Flaggy sandstone was also noted in a stream section [4326 3729] at Thornhill.

The largest of the Cuckfield Stone outliers is that to the north-east of Pock Hill. Milner (1923b, p. 285, pl. 22) recognized that this outlier (of his Upper Tunbridge Wells Sand) "is much more extensive than is shown on the old map". He recorded the junction of sand and Grinstead Clay in a pit [4420 3715] in Bushy Shaw; in the spinney [?4416 3658] at the south end of Woodcock Shaw and in the lane near Ship-yard Pett [4382 3634]. In spite of the numerous old pits there are now no sections in the Grinstead Clay.

Chandler's Farm–Lyewood Common–Motts Mill–Eridge Green–Boarshead. Lower Grinstead Clay, Cuckfield Stone and Upper Grinstead Clay have a wide outcrop on the Chandler's Farm–Lyewood Common Outlier. Although the Upper Grinstead Clay is incomplete each division is estimated to be about 25 ft (7·6 m) thick, and locally the Cuckfield Stone appears to be even thicker, possibly reaching 40 ft (12·2 m). Topley (1875, pp. 82, 379) described a calcareous grit, the Beech Green Stone (see p. 65), from this horizon: " . . . it [the Cuckfield Stone] occurs in some thickness, and large quantities have been there dug for roadstone. The grit is in two, or sometimes three layers, three feet thick together, and with it there is some sand." There are numerous old quarries at this horizon, but the pit referred to by Topley was probably one of several [492 377] on the east side of the road ½ mile (0·8 km) at 150° from Beechgreen

House. Many of the pits are now overgrown or partially infilled but several feet of thick calcareous sandstone can still be seen in two of them [4914 3766 and 4919 3754].

Thick sandstone floors the stream [4932 3781] in Minepit Shaw, and flaggy sandstone was exposed in sections in ditches [4932 3764, 4937 3759 and around 5020 3718] and streams [4743 3852, 4832 3824 and 4934 3796].

Road widening at Lyewood Common [4995 3709], although it gave no clear sections, showed that the Cuckfield Stone had thinned rapidly and only a couple of feet of sand and siltstone separated the Upper and Lower Grinstead Clay. No Cuckfield Stone is found south-east of this locality in the Grinstead Clay outcrops south of the River Medway. Some 600 yd (549 m) at 260° from Tye Farm the Cuckfield Stone locally disappears and the upper and lower clays are in contact [476 385].

The Lower Grinstead Clay outcrop is readily traceable by the number of old workings at this horizon. At only one point [5044 3718] was red clay present at the top, although in the Upper Grinstead Clay red clay was found at various levels [4787 3841, 4805 3835, 4877 3805 and 4920 3732].

South of the Medway the Grinstead Clay forms a number of disconnected and scattered outliers. In the railway cutting near Ham Bridge, Topley (1875, p. 89) recorded 25 to 30 ft (7·6–9·1 m) of clay and shale, with a reddened top, which he assigned to "Weald Clay". This is Grinstead Clay, and the sandstone above it is now classified as Upper Tunbridge Wells Sand (see p. 79). Red Clay was found at two points [511 364 and 5149 3650] on the Hunt's Farm Outlier, and could be traced continuously beneath the Upper Tunbridge Wells Sand along the top of Harrison's Rocks [5324 3587] to a point [5386 3555] 500 yd (457 m) S. of Park Corner. Isolated occurrences of red clay were noted near Cobbarn Farm [5411 3522] and in The Warren [5516 3624]. Red clay is also traceable along the top of the outcrop [533 349] to the north-east of Mott's Farm and in the small faulted outcrop [510 349] ½ mile (0·8 km) N. of Lye Green.

There are numerous old overgrown and flooded pits along these outcrops but no sections at the present day. Workings in Ligg's Wood [5335 3485] referred to by Milner (1923a, p. 50) as in Wadhurst Clay are in fact in Grinstead Clay.

A pit [540 332] has been dug through the clay to work the underlying Ardingly Sandstone at Boarshead but there are no exposures.

Bassett's Farm–Frienden Farm outliers. The Grinstead Clay and Cuckfield Stone crop out on the southern limb of the Finch Green Syncline. On both limbs the Cuckfield Stone appears as a wide expanse on the dip slopes. The Lower and Upper Grinstead Clay and the Cuckfield Stone are all about 20 ft (6·1 m) thick.

Both the Lower and Upper Grinstead Clay tracts are dotted with numerous old workings which enable their relatively narrow outcrops to be traced with ease between the unscarred Upper Tunbridge Wells Sand, Cuckfield Stone and Ardingly Sandstone. Springs occur at the base of both the Cuckfield Stone [5080 4098] and Upper Tunbridge Wells Sand [5074 4122]. Red clay was found everywhere at the top of the upper clay, but only locally beneath the Cuckfield Stone (e.g. around Blacklands Wood [510 413]). The only exposure of the Cuckfield Stone was at its base in the lane [5049 4092] leading to Hobbs Hill, where thin flaggy siltstone was noted.

Ashurst–Fordcombe–Rusthall–Tunbridge Wells. The Lower Grinstead Clay of the Ashurst Outlier is only about 15 to 20 ft (4·6–6·1 m) thick, and without exposure. Cuckfield Stone was formerly worked in the shaws [516 388 and 517 387] 800 yd (732 m) S.W. of Stone Cross, but now the only exposures are of flaggy sandstone seen at the roadside [5200 3905 and 5209 3905] and on the western side of the small fault at Stone Cross [5208 3892].

Around Fordcombe the Lower Grinstead Clay appears to maintain a fairly constant thickness of 15 to 20 ft (4·6–6·1 m). The Cuckfield Stone is here much siltier than at the outcrops so far described, and it is possible to trace its gradual thinning from a maximum of 15 to 20 ft (4·6–6·1 m) on the west side of the outlier to its disappearance in Priest Wood [5370 3957]. No reddened top to the Lower Grinstead Clay was found, but red clay was proved by augering at several points beneath the Upper Tunbridge Wells Sand [5291 3998, 5324 3988 and 5337 3997].

The Grinstead Clay was extensively worked in the outlier to the north of Langton Green. All the pits are now degraded or flooded.

To the south of the Tunbridge Wells Fault a narrow outcrop of Grinstead Clay, here some 15 to 20 ft (4·6–6·1 m) thick, can be traced in an east to west direction from its faulted junction with the Ashdown Beds at Ramslye Farm, around High Rocks to Top Hill. West of Top Hill the Grinstead Clay dies out. As with so many of the Grinstead Clay outcrops the bed can be readily traced by the number of old workings in the clay. Red clay was locally proved beneath the Upper Tunbridge Wells Sand [5372 3844, 5400 3859, 5491 3873, 5572 3856, 5614 3815, 567 380 and 5705 3785].

Grinstead Clay underlies much of Rusthall Common and clay was extracted in several large pits. The common is capped at its highest point by a small outlier of Upper Tunbridge Wells Sand. Red clay was proved at one point [5616 3926] beneath the sand.

A thickness of 17 ft (5·2 m) of Grinstead Clay was proved in the borehole at the former Culverden Brewery (see p. 81). Most of the outcrop is beneath the built-up part of the town of Tunbridge Wells, but to the north-west of the brewery site the clay can be readily traced. There are numerous pits at this horizon, some of which were in work until about the outbreak of war in 1939 (see Dines and others 1969, p. 47). A reddened top to the clay was found at two points [5780 4118 and 5797 4109].

The discontinuous lenticular silty clay to the south-east of Tunbridge Wells may represent the feather edge of the Grinstead Clay (see p. 83).

Farther south, to the west of Bell's Yew Green, the clay capping low crags of Ardingly Sandstone is the most easterly occurrence of Grinstead Clay which can be identified with certainty; it reaches a maximum thickness of about 15 ft (4·6 m). Several old pits pock its outcrop but there are no exposures.   C.R.B.

A site investigation borehole at Bartley Mill (see p. 84 and Appendix I) [631 656], 800 yd (732 m) at 196° from Wickhurst, proved 3 ft (0·9 m) of grey silt and silty shale resting on a megarippled surface of medium- to coarse-grained sandstone containing small pellets of siltstone and mudstone. This in turn rested on massive sandstones of Ardingly Sandstone type. It is therefore considered that the silt and shale may represent an attenuated equivalent of the Grinstead Clay resting on the Top Lower Tunbridge Wells Pebble Bed. If this interpretation is correct then this is the most easterly locality at which the Tunbridge Wells Sand can be subdivided in detail.   R.W.G.

Danehill–Sheffield Park–Maresfield. An account of the stratigraphy and nomenclature of the Lower Tunbridge Wells Sand (Ardingly Sandstone) and Grinstead Clay of the Danehill to Chailey area is given on p. 88. Between this area and Maresfield the Grinstead Clay maintains a fairly constant thickness of about 20 to 25 ft (6·1–7·6 m).

At Danehill an old quarry [4008 2746], 150 yd (137 m) W. of the church, shows 4 ft (1·2 m) of massive fine-grained, iron-stained sandstone overlain by 5 ft 6 in (1·7 m) of festoon-bedded sandstone of the same lithology. The massive appearance of the lower part of the section is probably due to lack of weathering. The lane adjacent to this quarry runs along a step between two marked sandstone features and, at the side of the field immediately below the lane, a thin clay bed was proved by augering. This clay was taken to be an attenuated Lower Grinstead Clay similar to that seen at Freshfield Lane Brickworks (see p. 88). The sandstones are therefore Cuckfield Stone. The small pond at the northern end of the quarry may be held up by this clay seam. Sandstones similar to the above occur in the stream course [3929 2760], 1050 yd (960 m) at 280° from the church, where they are overlain by grey and red clays of the Upper Grinstead Clay. Cuckfield Stone was probably dug for building stone from beneath a capping of Upper Grinstead Clay in a large old pit [3935 2754] adjacent to this stream.

The Grinstead Clay is about 25 ft (7·6 m) thick in the three inliers south-west and south of Danehurst [402 266], the top 10 ft (3·0 m) or so weathering to a red clay. This thickness is similar to that of the Upper Grinstead Clay at Freshfield Lane Brickworks (20 ft (6·1 m)) and at other nearby localities on the Horsham (302) Sheet (20 to 25 ft (6·1–7·6 m)). Furthermore, at all these localities red clays are confined to the Upper Grinstead Clay. This suggests that the Grinstead Clay in the three inliers all belongs to the upper part of the formation. Description of the underlying sandstone sequence, part of which may be Cuckfield Stone, is given on p. 88. The junction with the overlying Upper Tunbridge Wells Sand is marked by a good feature, and commonly by a spring line.

Red clays near the top of the formation crop out in the fields [399 266] 350 yd W. of Danehurst and [406 248] 500 yd (457 m) at 075° from Ketche's Farm. Yellow and grey clays are revealed in several of the old stone quarries (see p. 88) at the base of the formation and in old pits higher in the formation which were dug either for marl or for ironstone. Examples of this latter type of pit occur at Butchersbarn [3966 2638], 700 yd (640 m) at 255° from Danehurst; at Kidborough [3924 2606], 1250 yd (1·1 km) at 244° from Danehurst; in Knell Wood [3922 2550], 1450 yd (1·3 km) at 258° from Slider's Farm; in Pound Wood [4051 2522], 600 yd (549 m) S. of Slider's Farm; and in Huggett's Wood [4039 2590], 200 yd (183 m) N.W. of Slider's Farm.

Around Warr's Farm [393 226] Cuckfield Stone over 20 ft (6·1 m) thick is overlain by about 25 ft (7·6 m) of Upper Grinstead Clay. The position of the Lower Grinstead Clay here is marked only by a break in slope at the foot of the steep Cuckfield Stone feature. Westwards from here, on the Horsham (302) Sheet, a thin grey and red silty clay can be augered at this level.

An old quarry adjacent to the lane [3910 2281] 300 yd (274 m) N.W. of Warr's Farm shows 15 ft (4·6 m) of Cuckfield Stone consisting of deeply weathered flaggy and festoon-bedded sandstones. At the same stratigraphical level another old pit [3936 2281], 260 yd (238 m) E. of the first, shows 10 ft (3·0 m) of dark brown fissile festoon-bedded sandstone with lenses of calcareous sandstone. By contrast, a further quarry at the same level [3945 2281], 110 yd (100 m) E. of this second pit, shows 12 ft (3·7 m) of massive sandstone without calcareous beds. All these apparent variations in the lithology of the Cuckfield Stone are due to secondary weathering processes, the unweathered lithology consisting of sandstones with lenses of calcareous sandstone.

On the opposite side of the valley a roadside section [3914 2302] 380 yd (347 m) at 172° from Senates Farm shows 6 ft (1·8 m) of festoon-bedded dark brown sandstone which is probably Cuckfield Stone separated from the remainder of the outcrop by the Scaynes Hill Fault.

Cuckfield Stone has been worked from beneath a capping of Upper Grinstead Clay in a number of pits [3932 2260, 3937 2256 and 3960 2261] near Warr's Farm and near Senates Farm [3920 2325 and 3938 2326]. The last of these pits, 300 yd (274 m) at 110° from Senates Farm, is most unusual in that it is sited on the Upper Tunbridge Wells Sand, and must have penetrated the full thickness of the Upper Grinstead Clay to reach the Cuckfield Stone. This can only have been worth while if the stone here was of unusually good quality or if the overlying clay contained usable ironstone or limestone beds.

At Warr's Farm the basal few feet of the Upper Grinstead Clay are silty and there is no evidence that the Cuckfield Pebble Bed is present. Red clays occur at the surface south and east of the farm on the face of a good feature which is capped by the Upper

Tunbridge Wells Sand. In Warr's Wood [396 225], east of the farm, the top of the Grinstead Clay is marked by several springs. Red and grey clays also occur in the old pits near Senates Farm.　　　　　　　　　　　　　　　　　　　　　　　R.W.G.

The outcrop traceable from Furner's Green to Piltdown is extensively pitted, but the only exposure noted was of bluish grey clay in the lane [4128 2562] leading to Sheffield Mill. The junction with the Upper Tunbridge Wells Sand is commonly marked with a spring and red clay was noted at several points at the top of the Grinstead Clay, both on this outcrop [4278 2502, 4290 2537 and 4495 2365] and also on the outlier at Horney Common [4460 2594, 4505 2573 and 4558 2573].

The Grinstead Clay thins to about 10 to 15 ft (3·1–4·6 m) on the eastern side of the Horney Common outlier, around the A22/B2026 road junction, and on the High Hurstwood outcrops.

To the north of Uckfield the Grinstead Clay and Upper Tunbridge Wells Sand cap the Ardingly Sandstone. Within the present district the outcrop of the Grinstead Clay is difficult to trace through Shermanreed, Paygate and View woods, but on the adjacent Lewes (319) Sheet the continuance of this outcrop is readily mapped.　c.r.b.

UPPER TUNBRIDGE WELLS SAND

Dormans Park. A maximum of about 120 ft (36·6 m) of Upper Tunbridge Wells Sand, consisting of interbedded silts, silty sandstones and silty clays is exposed in this area.

A thin, but very persistent, slightly silty red clay occurs here and on the Horsham (302) Sheet at about 50 ft (15·2 m) above the base of the formation. A borehole adjacent to the railway [397 405] started at about the level of this red clay and proved 55 ft (16·8 m) of Upper Tunbridge Wells Sand resting on Grinstead Clay. About 5 ft (1·5 m) of red clay were formerly worked for brickmaking in a number of shallow pits [4025 4068] adjacent to Brickfield Wood, 700 yd (640 m) S. of Apsley Farm. Around Wilderwick [409 404] this clay is locally thicker and forms a wide outcrop. Another red silty clay, thinner and less persistent, occurs in the railway cutting [3981 4092] 400 yd (366 m) S.W. of Apsley Farm, at about 40 ft (12·2 m) higher in the succession.

The remainder of the sequence is poorly exposed but appears to consist predominantly of silts with thin beds of sandstone. Flaggy sandstones crop out in the stream bed [404 408] 500 yd (457 m) at 050° from Apsley Farm and in the railway cutting [3976 4102] 380 yd (347 m) at 250° from Apsley Farm. An exposure in the railway cutting [3980 4058] 350 yd (320 m) N. of the viaduct, just above the level of the persistent red clay shows: pale grey silts with purplish blotchy staining, 1 ft (0·3 m); compact, hard, iron-shot, silty cream-coloured fine-grained sandstone, one single thick bed, 1 ft 6 in (0·46 m); pale grey silt with some sand wisps, 1 ft (0·3 m). Both silt beds have a 'soapy' feel reminiscent of fuller's earth, a feature commonly encountered in this formation.

A section in the overflow of Cook's Pond [3971 4023], 80 yd (73 m) W. of the viaduct, which shows flaggy silty sandstones, silts and thickly bedded silty sandstones is close to the Blockfield Fault and must be very low in the formation.　　r.w.g., c.r.b.

Withyham–Eridge. Small outliers of silts and silty sandstone of the Upper Tunbridge Wells Sand cap the Grinstead Clay around Stoneland's Farm.

There are no exposures on the larger outlier around Park Corner but red and orange silty clays were commonly augered. A maximum of 75 ft (22·9 m) of strata are present. A trench section [5295 3480] across the outlier to the south of Leyswood (Lyeswood) revealed alternations of silts and silty sandstone.

Langton Green–Tunbridge Wells. To the south-west of Langton Green, owing to the absence of the Grinstead Clay, it is impossible to map separately the Upper and Lower Tunbridge Wells Sand. However the roadside section [5365 3881] of festoon-bedded sandstone is believed to be Upper Tunbridge Wells Sand. In a field [5442 3872] ¼ mile (0·4 km) S. of Langton Green a low rocky outcrop of massive sandstone lies immediately above the Grinstead Clay. Sandrock has not previously been noted at this horizon.

In the field [5546 3869] north-east of Holmewood House red silty clay was augered. Similar red silty clay, giving rise to poorly drained fields, crops out on either side of Coldbath Farm. Because of faulting the exact stratigraphical horizon at this locality is not known with certainty, but is believed to be within the Upper Tunbridge Wells Sand.

West of the main railway line a strong feature can be followed through the built-over portion of Tunbridge Wells. This is thought to be due to a hard sandstone band in the predominantly silty sequence. It was not possible to map this feature in the Hawkenbury area, nor in areas farther east around Hall's Hole and High Wood, where there is some uncertainty as to the correct horizon of the outcropping silts and sandstones (see pp. 83, 94). Red silty clays augered in the fields [around 5970 3805] west of Tutty's Farm may be the lateral equivalent of the clay seam traceable for a short distance on either side of the farm. Fine cream silty clay gives rise to poorly draining soil in Dunorlan Park [5975 3945]. Some 15 ft (4·6 m) of red and purple silt were noted in a temporary excavation [5776 3800] at Broadwater Down. The only exposures of sandstones were in the railway cuttings [5862 3880, 5918 3827 and 5929 3803].

There are no exposures in the outcrop to the north of Tunbridge Wells, which is largely built over.

Frant. East of the outcrop of Grinstead Clay in the grounds of Ely Grange it is difficult, in the absence of a diagnostic marker horizon (see p. 94), to assess how much of the outcropping sands and silts are referable to the Upper Tunbridge Wells Sand. The well [6052 3600] at the former Ware's Brewery penetrated a total of 165 ft (50·3 m) of strata before entering the Wadhurst Clay (Whitaker and Reid 1899, p. 36; Harvey and others 1964, 303/30). The Grinstead Clay and Lower Tunbridge Wells Sand together total about 130 ft (39·6 m) in this vicinity and so only the upper 35 ft (10·7 m) of 'stone and clay' can be assigned to the Upper Tunbridge Wells Sand.

Thin flaggy sandstone overlies thicker sandstone in the railway cutting [6092 3606] and farther south-east down the line thin flaggy sandstone and silts occur [6100 3574 and 6113 3566].

Sheffield Park–Fletching–Piltdown–Ringles Cross. The Upper Tunbridge Wells Sand is poorly exposed. Silt predominates and locally impersistent clay seams are developed. Soils are poorly draining, as in Wet Wood [406 231] and Great Wet Wood [413 226] and the silts and silty clays sufficiently impermeable to sustain large ponds such as those in Sheffield Park [417 240] and Piltdown Pond [444 223]. Sand and sandstone exposures are limited to old quarry sections or road cuttings. C.R.B.

The ridge running south from Danehurst and the adjacent ridge at Slider's Farm are capped by a bed of massive sandstone, about 20 ft (6·1 m) thick, which forms a strong feature. The base of this sandstone is about 20 ft (6·1 m) above the top of the Grinstead Clay, the intervening beds consisting largely of interbedded silts and silty sandstones. A small natural outcrop of massive greyish white sandstone [4042 2661] occurs on the crest of the feature in the grounds of Danehurst, 200 yd (183 m) E. of the house. An exposure showing 3 ft (0·9 m) of massive cross-bedded sandstone in the lane [4011 2477] adjacent to Ketche's Farm is within the same sandstone bed. Evidence from soil debris suggests that this sandstone is locally festoon bedded. R.W.G.

Flaggy siltstones and sandstones, inclined 5°N., can be seen by the roadside [4073 2633] opposite Heaven Farm. This northerly dip, possibly related to the Sheffield Forest Fault, is contrary to the regional dip of this outcrop which is about 2°S.

A clay seam, approximately 50 ft (15·2 m) above the base of the sand, has been mapped around Sheffield Park Farm. This is at a height similar to that mapped at Wilderwick (see p. 96) and Buckham Hill on the adjacent Lewes (319) Sheet. The clay is about 20 to 25 ft (6·1–7·6 m) thick in the region of the farm, but thins southwards and cannot be traced for more than ¼ mile (0·4 km) S. of this point. A silt or silty clay 25 ft (7·6 m) above the base of the Upper Tunbridge Wells Sand appears to have been worked in an old brick pit [4077 2477] 600 yd (549 m) N.W. of Sheffield Park Farm, which ceased production around 1910.

Thick flaggy sandstone is inclined 3°S.W. in the sunken lane [4270 2323] south-west of Fletching.   C.R.B.

At Wapsbourne Farm the Upper Tunbridge Wells Sand is folded into a shallow syncline adjacent to the Scaynes Hill Fault. A thin silty red clay, possibly at the same level as that occurring at Sheffield Park Farm, can be traced around the nose of this fold in the woodland north of Wapsbourne Farm. Clay in an old shallow pit [4005 2381] 480 yd (439 m) N. of the farm, was probably worked for brickmaking.

Between the Scaynes Hill and Fletching faults the strata dip locally northwards, contrary to the regional southerly dip. A road cutting [4038 2307] 600 yd (549 m) at 198° from Sheffield Bridge shows interbedded thickly bedded sandstones and silts dipping 3° at 012° and a temporary section in a pylon foundation [4032 2271] 400 yd (366 m) at 335° from Rotherfield Farm showed flaggy sandstones, silts and silty clays dipping 3° at 350°.

South of the Fletching Fault the Upper Tunbridge Wells Sand dips gently southwards and poor exposures of silts, silty sandstones and silty clays having this dip can be seen in the railway cutting [4051 2221] 300 yd (274 m) S. of Rotherfield Farm.   R.W.G.

Chapter VI

PLEISTOCENE AND RECENT

GENERAL ACCOUNT

At times during the Pleistocene glaciation small permanent snow- and ice-fields must have capped the higher parts of the district and over the whole area the ground must have been frozen to a great depth. Under such periglacial conditions erosional and depositional features were produced, only remnants of which are now preserved. The crags and tor-like outcrops of the Ardingly Sandstone (see Plates VI and VIIB) and parts of the Ashdown Beds (see Plate VIIA) were probably formed, and the cambering of these same beds and the valley bulging of the Purbeck Beds, Wadhurst Clay (see Plate VIIIA) and Grinstead Clay probably mainly took place during this period. The corresponding depositional features of the period, mainly soliflucted sludges and river deposits, are now represented by Head Deposits and by River Gravels. Later amelioration of the climate has allowed fluvial action to remove all but a few small patches of these deposits. The higher terraces of the River Medway are the only easily recognizable deposits of Pleistocene age within the district but the older parts of some of the patches of Head deposits probably date from the same period.   R.W.G.

RIVER DEVELOPMENT

The major watershed of the High Weald crosses the area from west to east, separating the headwaters of the River Medway and the River Teise catchment areas in the north from those of the River Ouse and the River Rother in the south (Figs. 2, 5). This watershed coincides closely with the Crowborough Anticline for much of its length, but in places apparent river capture has caused it to be displaced a little to the south of the actual anticline crest. Examples of this capture are given below and in Fig. 5.

Unfortunately no river deposits remain to afford conclusive evidence of such captures, but the examples given are cases where, with little regard for structure, the streams have taken courses contrary to the general trend in the area. In most instances the ground has been considerably eroded since the capture first took place, so that old stream beds are now only represented by cols in the ridge making up the watershed. Generally in this area the River Medway is increasing its headwaters at the expense of the River Ouse, and the River Teise in its turn has captured part of the Medway.

Apart from these examples of capture, the river systems appear to be well adjusted to the structure. As the main watershed follows the Crowborough Anticline, so in the south-east, the Rother flows eastward, closely following the axis of the synclinal area south of Stonegate Station [659 271]. Here not only has the syncline been accentuated by faulting but also erosion is at present operating on an east to west belt of soft Wadhurst Clay.

99

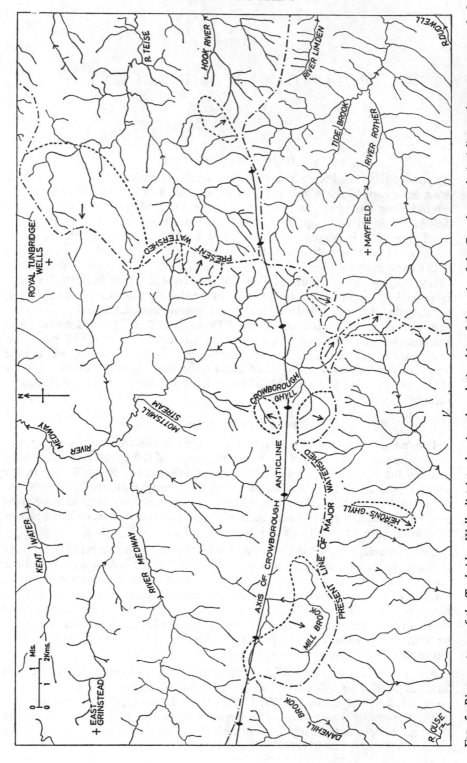

FIG. 5. River systems of the Tunbridge Wells district, showing the relationship of the axis of the Crowborough Anticline to the present watershed

The Medway flows north past Ashurst [512 390], again on a base of Wadhurst Clay. Some of its tributaries obviously follow fault lines, e.g. the tributary that flows through Groombridge [530 376]. In fact in many cases the tributaries of the four main rivers in the area are guided by the line of a fault for short distances before turning back to the main direction of drainage.

There is no evidence of a Mio-Pliocene peneplain in this area. Most streams are well adjusted to the structures and probably erosion by streams of the drainage pattern established since the Mid-Tertiary earth movements has continued with little interruption. Wooldridge and Linton (1955) suggested that this pattern is not generally due to epigenesis, but that in the south-west the Ouse might be epigenetic owing to submergence during the Pliocene era. Though this is possible, the large number of independent, immature sub-parallel streams in this area may primarily result from cutting mainly through sands, in contrast to other major rivers that have reached a base of Wadhurst Clay.

Such level surfaces as there are in this region vary greatly in height from place to place (Bird 1956; 1958) and are probably sub-aerial in origin. Benches recognized by Bird (1958) at 400 to 500 ft (121·9–152·4 m) and 250 to 350 ft (76·2–106·7 m) O.D. may well represent erosion during a period of relative standstill of the regional base level; further epeirogenic movements encouraged their later dissection to form unrelated units.

Most of the tributaries of the main rivers are cut in sands, silts and clays to form deep, wooded ravines. Only in sidestreams of the River Dudwell (a tributary of the Rother) has evidence been found of an old stream level, indicating later rejuvenation, about 15 ft (4·6 m) above the present stream base (Bazley and Bristow 1965). Where the rivers leave the area, however, they are generally more mature and flow through a widening belt of alluvium, forming meanders and sometimes ox-bow lakes. R.A.B.

HEAD (UNDIFFERENTIATED)

Sandy and silty downwash (mapped as Head) from hillsides capped by Tunbridge Wells Sand or Ashdown Beds is common. More local in occurrence are the spreads of Head on Wadhurst Clay at a distance from the Tunbridge Wells Sand outcrops. These latter deposits are usually associated with streams, and in these instances a certain amount of erosion and/or redeposition of an older Head or former terrace deposit has probably taken place. Where the Tunbridge Wells Sand provides the constituents of the Head, the upper topographical limit is usually the spring line issuing from the base of the Tunbridge Wells Sand. In certain valleys on the adjacent Sevenoaks (287) Sheet Head extends on to the floodplain of the Medway, where it merges with the Brickearth (Dines and others 1969, p. 122). The wash derived from a sandstone/clay junction makes the tracing of thin clay seams within the Tunbridge Wells Sand and Ashdown Beds particularly difficult.

In the east of the region Head was not more than 4 ft (1·2 m) thick over any area sufficiently large to represent on the map. Usually 3 ft (0·9 m) was the maximum. C.R.B., R.A.B., R.W.G., E.R.S.-T.

RIVER GRAVELS

River gravels, of limited extent within the present district, are confined to the valleys of the rivers Medway, Ouse, and a tributary of the Teise. It is only on the

Medway that a sequence of terraced river gravels can be recognized. In the region of Chafford Park [515 400] 2nd, 3rd and 4th Terraces are present, having respectively a base level approximately 10, 35 and 60 ft (3·0, 10·7, 18·3 m) above the river and corresponding to similar terraces on the adjacent Sevenoaks (287) Sheet. It is possible that a small area of 1st Terrace is present, beneath Head, at Ball's Green.

The terraced deposits of the River Ouse are assigned to the 1st Terrace and are partially obscured by Head.

In the east a small patch of gravel bordering a tributary of the River Teise is mapped as 2nd Terrace.

Pebbles of soft Wealden sandstone and siltstone are virtually the only constituents of the gravel, although occasional flint pebbles have been found in the Medway terraces (see p. 104). The deposits are of no commercial value but are dug at Chafford Park for domestic use.　　　　　　　　　　　　　　　C.R.B.

ALLUVIUM

Alluvium is present in all the major rivers and many of the minor stream valleys. The River Medway has the largest flood plain within the present district, reaching a maximum width of over 600 yd (549 m) at its confluence with the Kent Water near Ashurst.

All the rivers have their catchment on the Hastings Beds and in addition the River Dudwell drains a limited area of Purbeck Beds. The Alluvium of these rivers consists of fine sands, silts and clays of local origin. Occasionally lenses and seams of gravel are associated with the Alluvium. A maximum thickness of 28 ft (8·5 m) of clay, sand and stone has been recorded in the Alluvium of the Medway at Weir Wood, but thicknesses of 8 to 10 ft (2·4–3·1 m) are probably an average figure for the rivers Rother, Teise and Ouse. No deep buried channel was encountered beneath the Alluvium during the construction of Weir Wood Reservoir.

Many of the former Wealden ponds, established during the heyday of the iron industry, have now become silted up with an alluvial deposit.

Flooding is only serious in this district at Witherenden [654 268] where the alluvial plain widens at the confluence of several tributaries of the River Rother.　　　　　　　　　　　　　　　C.R.B., R.A.B.

CALCAREOUS TUFA

Small deposits of tufa are forming in several places where water saturated with carbonate comes to the surface, usually along limestone beds: near a spring from a calcareous horizon within the Wadhurst Clay in Stumletts Pit Wood [5287 2772], 160 yd (146 m) N.W. of Rumsden Farm, around a small spring issuing from the Blues Limestones of the Purbeck Beds, in the stream through Ten Acre Wood [6308 2193], about 1500 yd (1·4 km) E. of Tottingworth Park, and in many of the streams within the Purbeck Inlier.　　　　　　　　R.A.B.

DETAILS

RIVER DEVELOPMENT

The most obvious river capture is by the headwaters of the River Medway to the south of Wych Cross [420 318], where the present stream (Mill Brook) flows south and east for 2 miles (3·2 km) before turning sharply northwards to flow into the Medway. This stream is made up of some head streams originally of the River Ouse and now captured. The bulge of the watershed to the south indicates this capture. Similar river capture probably occurred near Stone Cross [518 287], Rotherhurst [553 285] and north of Argos Hill [568 283]. In each case capture of the headwaters of the Ouse by the Medway has pushed the watershed south of the crest of the Crowborough Anticline.

The north to south watershed between the Medway and its tributary the Teise in the north and the Ouse and the Rother in the south is also affected by river capture. North of Saxonbury Hill [578 330] capture by the Medway of part of the Teise headwaters appears to have taken place. But north of this, in the Hawkenbury [595 385] area of Tunbridge Wells, capture on a larger scale may have occurred, some 2 miles (3·2 km) of tributaries of the Medway now draining into the Teise.

Other examples of capture are to be found between the various tributaries of the same river, e.g. the headwaters of the Hook River (River Teise tributary) are probably captured by another Teise tributary to the north in the region of Pell Green [643 329]; at Crowborough [520 310] a tributary of the Medway has captured the headwater of a stream to the west and caused the zig-zag course of the present Crowborough Ghyll; below Heron's Ghyll [484 272] the peculiar course of the stream, possibly also affected by local faulting, is probably mostly due to capture of another tributary.   R.A.B.

HEAD (UNDIFFERENTIATED)

The lower boundary of the terraced gravels of the Medway near Walter's Green [515 405] and Willett's Farm [507 400] is obscured by downwashed gravel. Farther south, between Ashurst Station [507 388] and Burrswood [517 376], the Wadhurst Clay outcrop is obscured by Head, although locally pits have been opened along this tract, presumably to work the underlying Wadhurst Clay.

Near Boyles Farm [400 365], on a tributary of the upper Medway, fine brown sandy wash up to 6 ft (1·8 m) thick obscures part of the Wadhurst Clay outcrop adjacent to the Alluvium.

Temporary sections in the wood (Coneyburrow [602 410]) west of Gregg's Wood demonstrated that the Head was at least 7 ft (2·1 m) thick and overlay Wadhurst Clay. No Wadhurst Clay was exposed in the stream bed. The Head in Slowery Wood [410 646] and Sheepwash Shaw [648 412] to the east of Romford is over 4 ft (1·2 m) thick and is presumed to overlie Wadhurst Clay. It is only lower downstream on the adjacent Sevenoaks (287) Sheet that the clay is exposed in the stream bed.

Head overlies gravel at Ball's Green, and Topley (1875, p. 186) stated that at Summerford [498 367] "the old alluvium is chiefly loam"; and also noted that there was 6 ft (1·8 m) of brickearth in a railway cutting near Hartfield (op. cit., p. 187).   C.R.B.

Thick deposits of silty and sandy wash, largely derived from the Upper Tunbridge Wells Sand, flank the lower parts of the sides of the Ouse Valley. Since these deposits cannot be readily differentiated from the *in situ* weathering products of the Upper Tunbridge Wells Sand on which they rest they have not been completely delineated on the map.

A road cutting [4049 2338] 290 yd (265 m) at 200° from Sheffield Bridge shows 8 ft (2·4 m) of mottled orange and dark brown sandy wash. Adjacent to the alluvial flood-plain Head probably locally exceeds 10 ft (3 m) in thickness in this area. Near Wapsbourne Farm [399 234], on a tributary of the Ouse, sandy Head forms a gently sloping terrace-like feature about 10 ft (3·0 m) above the level of the stream alluvium. On the opposite side [399 231] of the same stream thick sandy wash obscures the outcrop of the Grinstead Clay.   R.W.G.

Five feet (1·5 m) of silty loam overlie the Upper Tunbridge Wells Sand in the sunken lane [4226 2283], 130 yd (119 m) S. of Mill Farm.

An extensive spread of Head borders the stream that flows through Buxted. At one point [4948 2359] 6 ft (1·8 m) of brown silty loam overlay 3 ft (0·9 m) of silt and silt-stone of the Lower Tunbridge Wells Sand.   C.R.B.

RIVER GRAVELS

RIVER MEDWAY

4th Terrace. Four small patches of gravel occur to the south, east and north-east of Chafford Park [516 394, 519 396, 5195 3985 and 5190 3995] with an estimated base level of 175 ft (53·3 m) O.D., some 60 ft (18·3 m) above the present river. There are now no exposures within this drift, but gravels have been worked in the past in shallow pits [5165 3940] to the south of Chafford Park.

3rd Terrace. Gravels of the 3rd Terrace have a distribution similar to those of the 4th Terrace. The base is about 150 ft (45·7 m) O.D., 35 ft (10·7 m) above the river. Only partially dissected, the deposits mainly form one continuous spread. A small pit [5153 3955] is worked just north of Chafford Park, the gravel being used for dressing the drives to the house. In 1963 the pit was 12 ft (3·7 m) deep and exposed 8 to 9 ft (2·4–2·7 m) of bedded, well-rounded and sub-rounded sandstone and siltstone pebbles, 1 to 3 in (25–76 mm) in diameter, beneath 2 to 3 ft (0·6–0·9 m) of brickearth. One flint pebble was found. The matrix consists mostly of small fragments of sandstone with very little interstitial sand (see Plate VIIIв). Older pits close by no longer expose the gravel.

The large gravel spread to the south of Willett's Farm rises to 175 ft (53·3 m) O.D. at the back part of the terrace and probably includes part of a 3rd Terrace, although the base close to the river is approximately 125 ft (38·1 m) O.D. Gravel lying at 30 to 40 ft (9·1–12·2 m) above the small stream, a tributary of the Medway, south-west of Redgate Mill [552 324] (Topley 1875, p. 187), may be assigned to the 3rd Terrace.

2nd Terrace. Gravel spreads of the 2nd Terrace occur on both sides of the River Medway in the region of Ashurst. The base of the gravel, about 10 ft (3·0 m) above the river, lies at about 140 ft (42·7 m) O.D. near Blackham Court [503 380]; this falls to about 125 ft (38·1 m) near Willett's Farm [507 400] and to approximately 115 ft (35·1 m) at Walter's Green [514 408].

An old gravel pit [5085 3975] 250 yd (229 m) S. of Willett's Farm exposed 6 ft (1·8 m) of gravel overlying the Wadhurst Clay. Most of the pebbles were of Wealden sand-stone and siltstone and were generally less than 2 in (50 mm) in diameter, with occa-sional large tabular masses of sandstone up to 6 in (152 mm) square. A little ironstone was present. There was little sorting and no bedding. Topley (1875, p. 186) recorded 10 ft (3·0 m) of gravel in the above pit and noted that, with the exception of two flint pebbles all the pebbles were of Wealden origin. The gravel showed a synclinal shape presumably due to periglacial action.

Topley (1875, p. 186) noted gravel in the railway cutting at Ham Bridge [about 510 370] and in a cutting to the east of Groombridge.

Gravel noted by Topley (1875, p. 186) at Ball's Green may belong to the 1st or 2nd Terrace, but it is everywhere covered by Head and the height of the base of the gravel is unknown. One to two feet (0·3–0·6 m) of gravel, resting on Lower Tunbridge Wells Sand, were observed in the small stream [5038 3639] ¼ mile (0·4 km) E. of Ball's Green.

Eight feet (2·4 m) of gravel were recorded in the railway cutting north-east of Hart-field and 14 ft (4·3 m) near Lines Farm [about 445 350]. At this latter locality the gravel was composed entirely of Wealden pebbles and locally contained thin seams of loam (op. cit., p. 186).

(A 10308)

A. Valley bulging in Wadhurst Clay, near Frant

PLATE VIII

B. Gravel of the 3rd Terrace of the River Medway overlain by brickearth, at Chafford Park Farm, near Fordcombe

(A 10314)

RIVER TEISE

2nd Terrace. A small tract of 2nd Terrace [680 395] falls within the present area near Tong Farm and is co-extensive with similar deposits on the adjacent Tenterden (304) Sheet. In augering generally a red, brown, yellow or white sandy loam is encountered. The maximum thickness proved was 3 ft 6 in (1·1 m) in a ditch section [6795 3954]. Shephard-Thorn and others (1966, p. 91) recorded 3 ft (0·9 m) of gravel overlain by brickearth in a ditch just beyond the borders of the present map area.

RIVER OUSE

1st Terrace. Low-lying spreads of gravel, partially covered by Head and Brickearth, border the Alluvium of the River Ouse near Fletching. There are no exposures of the terrace within the map area, but surface indications are that the constituents are all Wealden pebbles, in marked contrast to the higher-level terraces of the Ouse near Piltdown, just beyond the southern border of the sheet, where flints are common on the surface.   C.R.B.

ALLUVIUM

Much of the Alluvium of the middle reaches of the Kent Water is of very recent origin and largely related to the former iron industry. All the former ponds are wholly (Goudhurst Pond [4360 4045]) or partially (Furnace Pond [453 399]) silted up and/or drained by breaching of the dam (Mill Pond [442 400]). Downstream from the Furnace Pond there is a well established meander belt.

Crowstone, a mass of ferruginized sand and gravel, is present in the bed [4174 3838] of the unnamed stream near Shovelstrode Farm.

During the construction of Weir Wood Reservoir the full thickness of Alluvium and associated gravels was exposed in the cut-off trench. The details of the Alluvium varied across the trench, but in general consisted of 10 to 12 ft (3·0–3·7 m) of grey mottled clay resting on 10 ft (3·0 m) of gravel which in turn rested on the Ashdown Beds. The gravels appeared to be banked against a valley bulge (see Fig. 6 and p. 111). A section in a trial pit [4065 3530] recorded by Mr. S. C. A. Holmes showed:

	Ft	in	(m)
Soft brown clay becoming grey and blue, partly peaty ..	10	0	(3·05)
Dark soft clayey peat with wood, hazel nuts, etc.	2	6	(0·76)
Compact bedded well-worn sandstone gravel	5	6	(1·68)

The log of Borehole 2 [4057 3535] gave the following descending sequence of the drift deposits: mottled clay, 2 ft (0·61 m); dark brown clay, 2 ft 6 in (0·76 m); light clay, 1 ft 6 in (0·46 m); sandy clay, 6 ft (1·83 m); dark brown sand, 9 ft (2·74 m); sandy clay, 6 ft 6 in (1·98 m); stones, 1 ft (0·3 m), resting on Ashdown Beds.

Twelve feet (3·66 m) of Alluvium were proved in a borehole [4279 4242] near Forest Row Station.

Five feet (1·5 m) of mottled silty Alluvium overlie silt of the Ashdown Beds in the stream bed [5368 3760] 700 yd (640 m) E. of Groombridge.

A series of wells at Groombridge Pumping Station [528 365] were sited on the Alluvium of an unnamed stream to the south of Groombridge. Although the logs from these wells are not very precise they show that the Alluvium extends to a depth of at least 20 ft (6·1 m).

A small amount of crowstone can be found in the stream bed [4125 2651] 750 yd (686 m) S. of Tanyard Farm and also close to this stream's confluence [4180 2349] with the River Ouse to the west of Fletching.

The course of the Ouse has been much affected by straightening and canalization, although the river is now no longer used for navigational purposes within the present area. The remains of old locks can be seen 1000 yd (914 m) S.S.E. of Sheffield Bridge [4097 2275] and at Bacon Wish [3988 2408], 900 yd (823 m) at 300° from the Bridge (see Gibbs and Farrant 1970/71).

Straker (1931, p. 401) reproduced a 1724 map of the Maresfield Forge area. This demonstrated that the stream on the west side of Budlett's Common was navigable at that time. He also has an interesting aerial photograph (facing p. 402) of part of the stream which flows to the west of Maresfield and of which the above stream is a tributary. Prior to 1929 the former pond area [460 226 to 456 231] had silted up and/or become drained and was very boggy. In the same year the lower dam was repaired and the ponds refilled as far as the cross bay at Upper Forge [456 231]. By 1964 these ponds had reverted to a bog. Straker suggested that some of the higher meadows were also infilled or drained basins.   c.r.b.

In Jarvis Brook Brickworks [5300 2990], near Crowborough, the old course of the small nearby stream has been cut through during quarrying, exposing about 6 ft (1·8 m) of sandy wash and Alluvium above 1 ft 3 in (0·4 m) of gravel consisting mainly of sub-rounded locally derived Ashdown Beds sandstone pebbles.

A section [5135 2585] 450 yd (411 m) S.E. of the railway viaduct near Burnt Oak shows Alluvium consisting of fine sandy clay, 6ft (1·8 m); coarse gravel with sub-angular pebbles about 3 in (76 mm) diameter with much plant debris forming dark grey to black lenses, 6 in (152 mm); coarse gravel and subangular pebbles mainly of local Ashdown Beds sandstone and a log of wood 9 in (229 mm) in diameter, 1½ ft (0·46 m).

A typical section in the River Rother [5674 2688] about 1 mile (1·6 km) E. of Mayfield shows Alluvium composed of yellowish brown silty sand, 4 to 5 ft (1·22–1·52 m); gravel of poorly sorted, angular sandstone pebbles, with fine sand at base, 2 ft (0·61 m); mottled red and grey clay of Tunbridge Wells Sand, 2 ft (0·61 m).   r.a.b.

Chapter VII

SUPERFICIAL STRUCTURES

GENERAL ACCOUNT

IN THIS district four types of superficial structures affect the solid formations, i.e. landslipping, cambering, valley bulging and collapse depressions.

LANDSLIP

Landslips generally occur on the steep poorly drained slopes below the Lower Tunbridge Wells Sand/Wadhurst Clay junction. A similar relationship has been noted on the adjacent Sevenoaks (287) Sheet (Dines and others 1969). Slipping generally results from the failure of the weakened, water-saturated upper portion of the Wadhurst Clay. Although in a number of localities no major landslips were recorded on the clays of these slopes, it would require only a small change in the drainage or land-use pattern for some of these slopes to fail. Where trees are taken from clay slopes, landslipping often begins after a number of years when the roots have rotted away and lost their binding or stabilising effect. Such potentially unstable slopes occur near East Grinstead at Swite's Wood [402 397], near Danehill at Down Wood [394 273] and near the Sussex Oak Inn at Blackham [487 392].

The Purbeck Beds, particularly the upper clays, are also very liable to land-slipping and all around the deep-cut valley of the River Dudwell in the south-east corner of this area slips are common.

Weeks (1969) has recognized three main periods of solifluction near Seven-oaks. Lobes of soliflucted material extending from the Lower Greensand escarpment have been dated as Zone III of the Late-glacial sequence for the younger deposits overlying a more widespread sheet dating from the Weich-selian Glaciation. An older chert-laden solifluction sheet, capping Weald Clay hills, is ascribed to the Saale period of glaciation. By analogy it is presumed that some of the landslips within the present district commenced movement at the earliest of these three dates, although there is evidence for continuing movement at the present day (see pp. 109–10). The resulting topography is hummocky, often with lines of sedges marking the position of springs on the poorly drained ground.

CAMBERING AND VALLEY BULGING

The causes of cambering include sub-surface erosion and valleyward outflow of the underlying incompetent strata (Hollingworth and others 1944, p. 1). In the Tunbridge Wells district the Wadhurst Clay and occasionally parts of the Ashdown Beds are the incompetent strata. The valleyward flow usually takes the form of valley bulging, which is an expression of the differential loading of the

Lower Tunbridge Wells Sand and Wadhurst Clay of the valley sides compared to that of the Wadhurst Clay in the valley bottoms. The axes of the bulges tend to be parallel to the valley sides.

Cambering on a large scale often occurs on the Lower Tunbridge Wells Sand. Structure contours drawn on the base of this formation commonly show a valleyward dip resulting in apparent synclinal valleys and anticlinal ridges. This is well exhibited in the Langton Green area to the west of Tunbridge Wells (see Plate II), where the east to west 'anticlinal' axis approximately coincides with the Ashurst–Tunbridge Wells road, here following a ridge. There is a marked valleyward dip of the base of the Lower Tunbridge Wells Sand in a northerly, westerly and southerly direction. The westward dip into the valley of the River Medway is also mirrored by an eastward dip on the west side of that valley.

In some places it is evident that the valleyward dip of the Lower Tunbridge Wells Sand is at variance with the dip of the top of the Ashdown Beds. This was noted on the south-facing scarp of the Lower Tunbridge Wells Sand at Ashurst-wood [420 363].

The southern limb of the South Stonegate Syncline is probably accentuated by cambering.

The effects of cambering are more readily observed in the outcrops of Ardingly Sandstone and most exposures of this horizon show valleyward dips of 1° to 2°. The characteristic widened joints of the Ardingly Sandstone (see Plates VIA and VIIB) probably result from cambering.

In the Northampton area Boulder Clay, thought to be of Lowestoft age (Poole and others 1968, p. 52), overlies and clearly post-dates the origin of the cambers (Hollingworth and others 1944, p. 8), but ante-dates the formation of the river terraces in the valleys of the Nene and Welland (Kellaway and Taylor 1953, p. 358). Shotton (in discussion of Kellaway and Taylor 1953, p. 367) reached a similar conclusion on valley bulging affecting Pleistocene gravels in Yorkshire.

Worssam (1963, p. 129) considered that cambering in the Maidstone area may possibly have taken place under conditions of perennially frozen ground between the formation of the 4th and 1st terraces of the River Medway; he subsequently (1964b) suggested the possibility that there were two episodes of cambering, one dating from the Penultimate (i.e. Saale) Glaciation, the other from the last (Weichselian) Glaciation. Cambering that affects the Lower Greensand at the western end of the Weald (Thurrell and others 1968, pp. 95–7, 130–2) is considered to have taken place during the Saale Glaciation only. Valley bulges in some deeply cut valleys on the Weald Clay outcrop in that area are evidently of later date and may have been formed during Weichselian or even Post-Glacial times, during the latest phases of downcutting of the Arun Valley. While it is not possible to demonstrate that cambering is active today, there is evidence that manifestations of the related valley bulging, or similar phenomena artificially created by differential loading, still continue (Kent, and also Watson in discussion of Hollingworth and others 1944, pp. 35, 36; Kent in discussion of Kellaway and Taylor 1953, p. 372).

Reeves (1948, p. 258) noted landslips, valley bulging, cambering and gulls within the present district. One of the gulls, measuring 59 yd (54 m) long by 2 ft (0·6 m) deep, appeared in the Wadhurst Clay during wet weather. Earlier,

Topley (1875, p. 75) had figured what is now recognized as valley bulging in Wadhurst Clay near Mark Cross, but he attributed the contortions to faulting.

C.R.B., R.A.B., R.W.G., E.R.S.-T.

COLLAPSE DEPRESSIONS

Circular depressions formed by the collapse of the basal beds of the Wadhurst Clay into a cavity in the Ashdown Beds are present in Eridge Park [5702 3401]. Some 280 yd (256 m) at 075° from Hole Farm [5284 2279], Hadlow Down, the upper part of the Ashdown Beds has collapsed. These structures may have originated by subsidence into cavities formed along joint planes in the massive Top Ashdown Sandstone by the movement of underground water. In both the above cases the structures are circular and about 20 ft (6·1 m) in diameter. Another possible explanation is that these are collapses into old underground workings, but at these points no signs of any mine entrance were found. However, in Eridge Park, near the collapse structure mentioned above, there is a man-made tunnel, about 20 yd (18·3 m) long, leading to a circular cavern cut in sandstone. There is no known reason for mining at this horizon. In neither case was the depression likely to have been formed by shallow surface digging, such as bell-pitting, or by bombs dropped in World War II, although depressions caused by the latter have been reported.

Similar structures, but of relatively recent occurrence, have formed by collapse into the tunnels of the old ironstone mines in Snape Wood [6335 3021] near Wadhurst.

Shephard-Thorn and others (1966, p. 19) refer to oval or circular depressions in the Tenterden district formed by cambering on the interfluves of Wadhurst Clay.

Certain areas of high ground above camber slopes in the Maidstone district and south of Haslemere show ridge-and-hollow topography (Worssam 1963, pp. 127–9; *in* Thurrell and others 1968, p. 131) with surface depressions up to $\frac{3}{8}$ mile (0·6 km) long. However, there is no evidence for cambering under similar conditions in Eridge Park.

R.A.B.

DETAILS

LANDSLIP

Slipping has occurred 100 yd (91 m) E. of The Larches, where Ardingly Sandstone is faulted down against Wadhurst Clay [406 396].

Lower Tunbridge Wells Sand and Wadhurst Clay measuring 200 by 100 yd (183 by 91 m) has slipped down the eastern side of a small valley [445 402] 200 yd (183 m) E. of Scarletts. Farther east small slips of Lower Tunbridge Wells Sand and Wadhurst Clay are present on the eastern side of the valley north-east of The Moat [483 413 and 484 411], while small slips in the railway cutting [4853 4106] indicate instability where construction of the cutting has formed an outlier of Lower Tunbridge Wells Sand to the south-east of the main outcrop.

Small landslips of Wadhurst Clay measuring 200 by 100 yd (183 by 91 m) [498 387] and 200 by 60 yd (183 by 55 m) [496 384], are found to the north and south of the Lower Tunbridge Wells Sand outlier at Blackham. The outlier may be cambered towards the River Medway, as the base falls from about 325 to 150 ft (99·1–45·7 m) O.D. towards the river in less than $\frac{1}{2}$ mile (0·8 km), a general dip of about 4° (see p. 64 and Plate II).

At Danehill slipping of the Lewes Road [4055 2730] has been a problem for a number of years (Reeves 1948, p. 258). The ground falls from 381 ft (116·1 m) on the west side of the road to 150 ft (45·7 m) in the small valley to the east, a slope of about 8°. The fields east of the road are characteristically hummocky [4062 2728 and 4058 2716].

Slipping of clays within the Ashdown Beds took place in January 1962 on a new housing estate [6030 2235] at Broadoak. It appears that there was a down-dip slippage of a water-logged silty clay over a sandstone layer, caused by undercutting of the clay seam.

Small slips of Lower Tunbridge Wells Sand and Wadhurst Clay, with red clay at the top, occur in the railway cutting 400 yd (366 m) S. of Buxted [4960 2293]. There is evidence for movement of the Wadhurst Clay in the field [5315 3270] 400 yd (366 m) W. of Boarshead.

A small landslip involving the Wadhurst Clay was noted in 1963, but not mapped separately, in the fields [587 356] immediately west of the Albert Memorial, Frant. A reinvestigation of this area in 1970 demonstrated that this slip is a part of a much larger one at the junction of the Lower Tunbridge Wells Sand and Wadhurst Clay to the west of the Tunbridge Wells–Eastbourne road. The wooded nature of the ground, the thick undergrowth and partial obscuration by downwash make an exact delineation of the slip difficult. It can be traced southwards from the Memorial, parallel to the main road, to a point [5860 3459] just south-west of Sleeches Cross; westwards and south-westwards the slip extends for a ¼ mile (0·4 km) from the Memorial [to 5830 3535]. In addition to the hummocky terrain there are large recent cracks and tilted trees 270 yd (247 m) W. of the Memorial. In tracing the rapid southerly fall in base level of the Lower Tunbridge Wells Sand, a further area of suspected slippage, some 500 by 120 yd (457 by 110 m), has since been delimited tentatively 1000 yd (914 m) W.S.W. of the Memorial. It is possible, however, that this is a normal boundary, the unusual valleyward dip being accentuated by cambering. Immediately west of this suspected slip the junction of the Wadhurst Clay and Lower Tunbridge Wells Sand is obscured by downwashed sands and silts.   C.R.B.

Slipping of the Wadhurst Clay was evident in sections cut during the road widening scheme at Butcher's Cross [560 250] in 1965 (Bazley and Bristow 1965, p. 318). Another example of slipped Wadhurst Clay is the hillside north-west of Bestbeech Hill [6190 3167] which is continually moving, as evidenced by the local road surface.

Extensive slipped ground occurs along the Tunbridge Wells Sand–Wadhurst Clay junction west of Old Swan Farm [6521 3877], Lamberhurst Quarter, where it occurs on the steep eastern flank of a minor tributary valley of the Teise. Smaller slips in the same stratigraphic situation occur in the Teise Valley near Mount Pleasant at [658 371], and at [664 365] in Timberlog Wood.

In the Bewl Valley Wadhurst Clay inlier, a slipped area roughly 200 yd (183 m) square, affecting beds just at and below the junction with the Tunbridge Wells Sand, occurs on a steep north-west facing slope [661 323], ½ mile (0·8 km) at 300° from Chesson's Farm.

Strata high in the Ashdown Beds have been involved in a minor slip on a steep valley side [671 312], 400 yd (366 m) at 110° from Birchett's Green.   E.R.S.-T.

The Purbeck Beds, particularly the upper clays, are also liable to landslipping and all around the deeply cut valley of the River Dudwell in the south-east corner of this area slips are very common.   R.A.B.

VALLEY BULGING

A stream section east of Herontye shows several small exposures [4003 3671, 4017 3697 and 4027 3710] of contorted grey shales with an included thin Tilgate Stone bed. The fold axes strike almost exclusively N.N.E., parallel to the stream course.   R.W.G.

Contorted '*Cyrena*' shales were noted in the bed of the stream [4079 3405] which crosses the faulted outlier of Wadhurst Clay near Charlwood.

In the valley ½ mile (0·8 km) N.E. of Bayham Abbey the Lower Tunbridge Wells Sand–Wadhurst Clay boundary, partially obscured under hillwash, can be traced along the valley sides and at [6500 3747] crosses the stream. A few yards upstream red shaly clay, the uppermost Wadhurst Clay, is brought to the surface as a small inlier by valley bulging. Contorted clays and shales were noted in the stream [6222 3983] 500 yd (457 m) N.N.W. of Great Bayhall and along the stream to the east of the same locality [between 6304 3910 and 6327 3941], affecting limestones, ironstones, siltstones and shales of the Wadhurst Clay.

One of the best examples of valley bulging was that exposed in the cut-off trench during the construction of Weir Wood Reservoir, Forest Row [406 353]. The trench was cut across the flood plain of the River Medway, through the superficial alluvium and river gravels, some 20 ft (6·1 m) thick, into the Ashdown Beds. The general relationships of the beds can be seen in Fig. 6. The river deposits, sand and sandy gravel appear to be bedded against the bulge on the south side and not flexured with it. The bulge dies out at a depth of 60 ft (18·3 m) from the surface. There appears to be no evidence of movement in the strata adjacent to the bulge except possibly the thin 'band of soft broken mudstone' at a depth of about 50 ft (15·2 m), over which the higher beds may have slid. A similar bulge in Ashdown Beds was exposed in the cut-off trench of the Powder Mill dam on the adjacent Hastings (320) Sheet (Waters 1962, p. 59).   C.R.B.

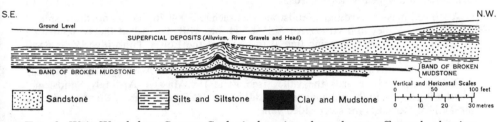

S.E.   N.W.

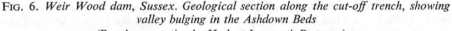

| | Sandstone | | Silts and Siltstone | | Clay and Mudstone |

FIG. 6. *Weir Wood dam, Sussex. Geological section along the cut-off trench, showing valley bulging in the Ashdown Beds*
(Based on a section by Herbert Lapworth Partners.)

Valley bulging and associated cambering have been revealed by borings on the site of a dam for a proposed reservoir in the Bewl Valley, 1½ miles (2·4 km) S. of Lamberhurst, which straddles the junction of this and the adjoining one-inch sheet. (Shephard-Thorn and others 1966, pp. 20–1).   E.R.S.-T.

In the valley of the Danehill Brook [3922 2734], 1120 yd (1024 m) W.S.W. of Danehill church, small exposures of vertical grey shales of the Wadhurst Clay were recorded.
                                                                                                   R.W.G.

Three valley-bulge structures in thin silts and sandstones of the Ashdown Beds were noted in the streams east of Heron's Ghyll [4862 2750, 4862 2726 and 4923 2740].
                                                                                                   C.R.B.

Stream sections close to an outcrop of Top Ashdown Sandstone [5560 2635], 700 yd (640 m) S.W. of Horleigh Green Farm, show contorted thin silts and sandstone. A small anticline in Ashdown Beds with its axis parallel to the valley sides was exposed in the bed of the River Rother just north of Lymley Wood [5677 2743]. Similar structures were noted in the stream just east of Downford [5767 2717]. Contorted silts and sandstones of the Ashdown Beds were again observed in the main stream at Coggins Mill [597 279] and for at least 800 yd (732 m) upstream. Again, as in the Lymley Wood section, the axes of the anticlines trend parallel to the valley sides. Similar disturbances in the Ashdown Beds were seen in the stream [5672 2984] 400 yd (366 m) N.W. of Yewtree Farm.

A stream section [6047 3495] about 1 mile (1·6 km) E.S.E. of Frant showed tightly folded and in places overturned shaly Wadhurst Clay (see p. 70 and Plate VIIIA).

Some 200 yd (183 m) N. of Fright Farm (now Saxonbury Farm) contorted medium grey mudstones of the Wadhurst Clay crop out in the stream bed (see Topley 1875, fig. 10, p. 75). Highly disturbed mudstones with thin nodular ironstone and '*Cyrena*' limestone crop out along the stream through Almond's Wood [5662 2515], about 870 yd (796 m) S.E. from Butcher's Cross. To the south of Mayfield, 360 yd (329 m) S. of Cranesden, valley-bulged grey mudstones were noted in a small deep-cut stream [5893 2633]. Contorted mudstones, limestones and siltstones 940 yd (860 m) E.N.E. of Button's Farm, south of Five Ashes, may be due to valley bulging or to faulting.

The Tunbridge Wells Sand is affected by valley bulging in the stream 600 yd (549 m) N.W. of The Mount, Wadhurst.

Valley bulging is a common feature in the River Dudwell where it cuts deeply into the dominantly clayey Purbeck Beds with the Ashdown Beds sandstone above.   R.A.B.

CAMBERING

At East Grinstead a small pit [3997 3841] 200 yd (183 m) S. of East Court, showed cambering in a sandstone bed in the lower part of the Lower Tunbridge Wells Sand. In the road cutting nearby at Blackwell Hollow [398 385] Ardingly Sandstone is cambered to dip locally northwards.   R.W.G.

A slipped mass of Ardingly Sandstone was seen [5020 4076] on the northern side of the Medway Valley 100 yd (91 m) N.E. of Hobbs Hill. Disturbed flaggy sandstone crops out just above the spring line at the base of the Lower Tunbridge Wells Sand [532 411] 200 yd (183 m) E.N.E. of Palmers. As mentioned on p. 109 the Lower Tunbridge Wells Sand outcrops on either side of the Medway near Ashurst [510 390] appear to have cambered towards the valley (see Plate II). Massive Ardingly Sandstone, weathering flaggy, in the old quarry [522 382] 500 yd (457 m) N.N.E. of Burrswood, dips 30° W.N.W. towards the steep-sided valley.

In the lane 700 yd (640 m) W. of Gilridge Farm a detached portion of Lower Tunbridge Wells Sand [5103 3287] is separated either by cambering or landslip from the main crop 30 yd (27 m) N. Farther west along the same outcrop cambering may account for the apparent thinning of the Wadhurst Clay to less than 100 ft (30·5 m) in thickness to the south and east of Gilridge Farm in Rough and Bream woods [520 327 and 523 328]. The general strike of the strata in this vicinity is approximately east to west, but both ground level and base level of the Lower Tunbridge Wells Sand fall from 530 ft (161·5 m) O.D. at a point 300 yd (274 m) W. of Gilridge Farm to 300 ft (91·4 m) O.D. to the east and west of this locality. To the east of Gilridge Farm [522 328] the base of the Lower Tunbridge Wells Sand is within about 80 ft (24·4 m) of the top of the Ashdown Beds.   C.R.B.

Cambering was clearly seen affecting sandstones in a sandstone/silt sequence at the Jarvis Brook Brickworks [5302 2980] (Bazley and Bristow 1965, p. 317). Similarly in the lane above Redgate Mill Farm, about 1½ miles (2·4 km) S. of Eridge, the steep dip down the hill of Ashdown Beds sandstones [5550 3252] is probably due to cambering.   R.A.B.

Chapter VIII

ECONOMIC GEOLOGY

AGRICULTURE AND SOILS

ON THE Purbeck inliers the stiff clay soils with limestone debris are lightened to some extent by a wash of sand from the surrounding slopes of the sands of the Ashdown Beds. Most of the ground in this area is situated on steep slopes and even with the wash tends to be very wet, with landslipping a common nuisance where unwooded. The Ashdown Beds contain a high proportion of fine-grained sand and silt and give rise to lighter, more loamy soil than do the Purbeck Beds, although locally the presence of the clay and silty clay seams results in heavier ground when wet. A typical view of scenery associated with the Ashdown Beds is shown in Plate I. The Wadhurst Clay makes a very heavy clay soil. However, the outcrop is often obscured by downwashed sands, particularly the beds close to the junction with the Lower Tunbridge Wells Sand (see p. 101) and in the valley bottoms. A deep soil cover on the lower slope of the valley sides is very noticeable in those areas where it is, or has been, standard practice to leave the ground bare between hop vines or trees. Rapid erosion of the soil takes place during periods of quite moderate rainfall. The Lower Tunbridge Wells Sand is, in general, coarser grained than the Ashdown Beds (see Dines and others 1969, p. 35) and interbedded clays are rare. The soils developed on this formation are light and free draining (see Plate VIB for typical Lower Tunbridge Wells Sand scenery). Those on the finer grained Upper Tunbridge Wells Sand are not so light as on the Ashdown Beds and interbedded thin silty clay seams locally give rise to heavier soils. C.R.B., R.A.B.

WEALDEN IRON INDUSTRY

A brief résumé of the industry is given by Shephard-Thorn and others (1966, pp. 106–7). Fuller details of the method of working, production figures, history of the forges and furnaces, etc. are contained in Straker's excellent work entitled 'Wealden Iron' (1931). Within the present area the main 'pay' horizon appears to have been the nodular or tabular bed of clay ironstone some 20 to 30 ft (6·1–9·1 m) above the base of the Wadhurst Clay (see p. 42), which was worked by bell-pitting. The location of the bell-pits has been recorded in the details of the Wadhurst Clay. Locally it has been possible to depict the outcrop of this ironstone seam on the six-inch and one-inch maps.

The final decline and extinction of the industry was thought by Straker (1931, pp. 65–6) to be due to the high cost of transport over poor roads which could only be used at certain times of the year, coupled with a remarkable dryness of climate in the first half of the eighteenth century; many of the streams appeared to have temporarily dried up and could not be used as a source of power. He stated, however, that the most important factor leading to the decline was the high cost of charcoal fuel. In the western Weald, Worssam (1964a, p. 540) believed that a factor in the decline of the industry was the depletion of iron ore reserves, although earlier Schubert (1957, p. 33), Sweeting (1944, p. 5) and Topley (1875, p. 332), referring to the central Weald, had not considered this factor to be significant. Deforestation of the Weald by the iron workers, often quoted as the main reason for the death of the industry (see Straker 1931, p. 109) appears not to have taken place. Indeed coppicing of the woods in rotation to provide charcoal, known to have been in use in 1581 (op. cit., p. 123), and in general practice by the mid-seventeenth century (Schubert 1957, p. 222) led to the conservation of woodland.

An attempt in 1857 to re-establish the industry by mining iron ore in the Ashdown Beds at Snape Wood [6335 3021], near Wadhurst, was abandoned the following year (see p. 59). The Earl of Sheffield clearly had hopes of reviving the industry on his estate when he commissioned Professor Gregory in 1908 to investigate "the prospects of Iron Mining and Smelting in the neighbourhood of The Sheffield Park, Sussex". Gregory's concluding remarks are still relevant today, "so thin a vein . . . could not be worked at a profit under any conditions of costs and prices likely to arrive in the near future" (Gregory 1908, p. 22).   C.R.B.

BRICK CLAY

Brick-making is of limited importance in the Tunbridge Wells area. Tiles have been found that were made in the Wadhurst district during the Roman occupation and there have probably been numerous small domestic works between that period and the present day. Within the district there is only one working pit, the Jarvis Brook Brickworks [5310 2970], near Crowborough. The bricks are made from light grey silts, with only thin clay seams, that occur in the middle of the Ashdown Beds.

Although there are now no working pits in the Wadhurst Clay, the numerous pits dug into this formation probably witness extensive earlier use for brick-making; many of them, however, were dug for marl and/or ironstone (Topley 1875, p. 335). A large proportion were opened close to, or at, the Lower Tunbridge Wells Sand/Wadhurst Clay boundary, presumably to exploit both formations, as is or was done at the Quarry Hill and High Brooms Brick and Tile Company Limited pits on the adjacent Sevenoaks (287) Sheet (Dines and others 1969, p. 145). A brickworks [451 342] using basal Wadhurst Clay at Upper Hartfield appears to have been in production until about 1960 (see p. 67). It is possible that the basal Wadhurst Clay and upper Ashdown Beds were worked in a manner similar to the Tunbridge Wells Sand and Wadhurst Clay exploitation. One such location is near Yewtree Farm [5718 2978].

Clay for brick and tiles was probably once dug from the top of the Purbeck Beds near Poundsford [6390 2230], south of Burwash Common.   C.R.B., R.A.B.

CEMENT AND LIME

Cement and lime were once made from the limestones of the Purbeck Beds. The Greys Limestones and Blues Limestones have been dug extensively in bell-pits in the inlier in the south-east of the area. The many bell-pits along the outcrop at these horizons indicate that the beds are almost worked out. The last mining was early in the nineteenth century (see Gould *in* Topley 1875, pp. 384–6). The Blues Limestones are reputed to have produced lime for building purposes superior to the Greys Limestones, but both make equally good land-dressing.   R.A.B.

BUILDING STONE

The shelly limestones from the Greys Limestones of the Purbeck Beds have been occasionally used for building stone. These were dug around the Purbeck inlier in the south-east. Examples can be seen in the garden of Bateman's [6710 2379], near Burwash, where they are used as paving stones. The house at Bateman's, once the home of Rudyard Kipling and now National Trust property, is constructed of local Ashdown Beds sandstone probably obtained from a small quarry less than 100 yd (91 m) away (see Geological Survey Photograph A 10144).

Many stone workings are restricted either to the Ardingly Sandstone or to the top Ashdown Beds where overlain by clay. The advantages of working the beds at these horizons are, firstly, that the large blocks, devoid of minor jointing, enable the stone to be worked in any direction, and, secondly, that the stone is soft when first sawn and the surface hardens on exposure (Topley 1875, p. 367). There are now no quarries operating within the present district, but the still active Philpots Quarry, West Hoathly, on the adjacent Horsham (302) Sheet, provides a good example of how the stone is worked. A facies similar to the Ardingly Sandstone occurs some 60 ft (18·3 m) above the base of the Lower Tunbridge Wells Sand in the eastern part of the area; this has also been worked for building stone. The locations of many of the former pits are given in the details. Topley (1875, p. 367) referred to the building stone from Calverley Quarry, Tunbridge Wells. Bayham Abbey near Lamberhurst (Plate IX) provides a good example of the use of the Ashdown Beds as a local building stone.

C.R.B., R.A.B.

SAND AND GRAVEL

Sand has been obtained in the past particularly from the Top Ashdown Sandstone and the massive sandstones within the Tunbridge Wells Sand. There are many old sandpits at these horizons all over the area. Again no large commercial organisations are utilising this source of sand at present, probably because the sand is too fine for most purposes.

Topley (1875, p. 395) mentions the use of glass sand from the Hastings Beds, but gives no details. Milner (1923b, p. 288; 1924, p. 386) refers to "glass-sand" facies of the Lower Tunbridge Wells Sand, but does not cite examples of the sand being worked for this purpose. Boswell (1918, p. 55) gives an example of pits in the Tunbridge Wells Sand (Ardingly Sandstone), at Ashurstwood, suitable for glass-sand extraction, but does not make it clear whether the sand has ever been used for this purpose. There appears to be no historically early Wealden Glass industry in this district comparable to that which existed in the western Weald during the Middle Ages (Kenyon 1967).

The only gravels within the district are composed largely of material of Wealden origin. Gravel has been dug in the past from pits bordering the Medway to the north of Ashurst. There is a small pit at Chafford Park Farm, where gravel is still worked for domestic purposes (see p. 104 and Plate VIIIB).

C.R.B., R.A.B.

ROADSTONE

At present the only stone used commonly for hardcore is the limestone produced as a by-product of the gypsum mining in the Purbeck Beds at Mountfield and Brightling, in the south-east corner of the map area. This hardcore is only used locally.

Gravel of Wealden debris is locally used on tracks but is unimportant.

Cinder (= slag) from the old ironworks has been widely used as roadstone. The Roman London–Lewes Way, averaging 12 to 15 ft (3·7–4·6 m) wide, has a metalling of cinder of 12 to 15 in (0·30–0·38 m) thick for most of its route across the Hastings Beds (see frontispiece in Margary 1948). The volume of cinder was enormous and is evidence of a very extensive industry at this time. Straker and Margary (1938, p. 58) estimated that 500 000 cu ft (45 000 cu m) of cinder, weighing some 35 000 tons, were used on this road alone. Topley (1875, p. 379) recorded that cinder was one of the best road materials within the Weald.

Sandstone from the Ashdown Beds was used by the Romans for metalling the Lewes–Brighton Way across Ashdown Forest (Margary 1948, p. 143). The 'Crowborough Stone' (Ashdown Beds) has been used for a similar purpose in more recent times (Topley 1875, p. 379).

The Beech Green Stone (Cuckfield Stone, see p. 92), a calcareous grit, was formerly dug in large quantities from pits [about 492 376] to the north of Withyham.   C.R.B., R.A.B.

GYPSUM

Gypsum from the Lower Purbeck is at present being worked at the Mountfield Mine on the Hastings (320) Sheet and at the Brightling Mine [6770 2187] just on the border of this sheet. The gypsiferous beds in the Brightling area are about 50 ft (15·2 m) thick; the gypsum is in seams varying from 4 ft 6 in to 14 ft (1·4–4·3 m) thick, with shale partings.

In 1968 the Mountfield Mine (Anderson and Bazley 1971) was producing some 200 000 tons per annum, all of which is extracted from the lowest (No. 4) seam; production in the upper beds has now ceased. The bulk of the former production has since 1963 been transferred to the Brightling Mine, now producing some 450 000 tons per annum, which is made up of 180 000 tons per annum from the top No. 1 seam, approximately 6 ft (1·8 m) thick, and 270 000 tons from the basal No. 4 seam.

The system of working is identical at both mines, i.e. pillar and stall with inter-seam drifts. Since 1965, following an emission of methane gas in No. 1 seam, the Brightling Mine has been declared a 'safety light' mine.   R.A.B.

OIL AND GAS

The Jurassic rocks underlying the central Weald have been known for many years to contain traces of oil and gas. Several formations, including the Lias, Kimmeridge Clay and Purbeck Beds, are known from other areas to be potential source rocks, but these formations are by themselves generally too argillaceous to permit direct extraction. Exploration has therefore been aimed at finding reservoir-rocks, for example in the Inferior Oolite, Great Oolite, Corallian Beds, Portland Beds and Ashdown Beds, wherever thought to be suitably folded or

(A 10146)

REMAINS OF BAYHAM ABBEY, NEAR LAMBERHURST

PLATE IX

faulted. It is known from borehole evidence that many of the surface structures in the Weald continue at depth and that they may well reflect older structures in the sub-Mesozoic floor. For this reason geophysical work, designed to plot the hidden extensions of the more important surface structures, has been used extensively in siting exploratory boreholes for oil and natural gas.

So far the results have been disappointing. In 1895 and 1896 natural gas was encountered in two boreholes at Heathfield [5803 2135] which probably passed into Upper Purbeck Beds. The gas from the second borehole was used to light the railway station until 1964 when the railway was closed. Further exploration in the area was unsuccessful and it was concluded that the gas field was a small local pocket which had collected in an anticlinal crest. In 1955 a nearby well [5859 2150] found small supplies of gas (Falcon and Kent 1960). A borehole near Brightling [6726 2182] also produced gas and a little oil, but below commercial level (op. cit.). The Ashdown Nos. 1 and 2 boreholes, the most important exploratory drilling in the district, were sunk to test the Crowborough Anticline. Although oil traces were found at several levels in the Jurassic rocks, commercial quantities were not present.   R.W.G., R.A.B.

WATER SUPPLY

The major surface water divide between the areas draining northwards to the River Medway and those draining southwards to the rivers Ouse and Rother crosses the district in an approximately east to west direction, from 2 miles (3·2 km) east of Wadhurst to a point 3 miles (4·8 km) south of East Grinstead. For 1½ miles (2·4 km) in the west and 6 miles (9·7 km) in the east of the district the divide coincides closely with the generally east to west trending axis of the Crowborough Anticline (Fig. 5). In the central tract, however, the surface divide is up to 1½ miles (2·4 km) to the south of the anticlinal axis. Measurements of the groundwater level in the west of Ashdown Forest and around Rotherfield suggest that the groundwater divide in those areas coincides with the anticlinal axis and not with the surface divide. There is no evidence in the intervening ground, but the possibility cannot be discounted that groundwater may drain south into the Sussex Ouse from some 7 square miles (18 sq km) of the Medway surface catchment, lying to the south of the Crowborough Anticline.

Earlier references to the water supply of this area may be found in Topley (1875), Whitaker and Reid (1899), Whitaker (1908; 1911; 1912), Edmunds (1928) and Buchan and others (1940). A Well Catalogue (Harvey and others 1964) has included abstracts and information on virtually all recorded wells in the district.

The total groundwater abstraction from the district in 1964 amounted to 646·2 million gallons (m.g.) according to returns received under Section 6 of the 1945 Water Act. Of this, 607 m.g. (1·8 m.g. per day) were taken from six sites in the Ashdown Beds, all for public supply. The remaining 39·2 m.g. were abstracted from two sites exploiting Tunbridge Wells Sand, 0·2 m.g. of this for industry, the remainder for public supply. Assuming that additional, unrecorded, abstraction is equal to 10 per cent of that recorded, some 2 m.g.d. is probably abstracted from underground sources. In this district, water is known by the Kent River Authority to be abstracted intermittently from the River Medway at ten localities. The abstraction did not exceed 0·03 m.g.d. in 1964.

The River Medway has been gauged continuously at Chafford [517 405] by the Kent River Authority since October 1960 (Anon. 1962). The catchment area to the gauge is 98·5 square miles (252 sq km), of which all but some 10 square miles (25 sq km) lie within the present district. Geologically the catchment consists of 60 square miles (153 sq km) of Ashdown Beds, 19·0 square miles (49 sq km) of Tunbridge Wells Sand and 19·5 square miles (50 sq km) of clays (Wadhurst Clay, and clays within the Tunbridge Wells Sand). The groundwater contribution to total discharge has been determined by analysis of the stream hydrographs, and amounts to an average of 6·6 in (167 mm) over the catchment, for the period 1st October 1960 to 31st September 1965. This figure is 45 per cent of the average total discharge for the same period. No correction has been applied to allow for water used to supply Crawley from the Weir Wood reservoir in the upper reaches of the Medway, but the effect of this on the groundwater contribution to total discharge is thought to be slight.

The average figure of 6·6 in (167 mm) for groundwater discharge can be used as an estimate of infiltration into the principal aquifers. For the Ashdown Beds it will represent a minimum figure because the groundwater discharge as estimated from the hydrographs will include only limited contributions from the Tunbridge Wells Sand and from the clays. The Tunbridge Wells Sand probably, and the clays certainly, contribute less groundwater to river discharge per square mile of their outcrop than do the Ashdown Beds. Over the Ashdown Beds in this district an infiltration of 6·6 in (167 mm) is equivalent to 0·3 m.g.d. per square mile.

The percentage of groundwater in total river discharge is on average greatest during the summer months May to August inclusive, when rainfall is normally low and evaporation is at its highest. Increments to the groundwater zone can occur at any season, however, depending on soil moisture content and on the intensity and duration of rainfall and other meteorological factors. The discharge data for the River Medway indicate that in the present district the highest proportion of groundwater in river discharge occurs on average in July (74 per cent) and the lowest in November (41 per cent).

Drift. No wells are known from which groundwater is at present being taken from superficial deposits although in the past some domestic supplies were obtained from them. In fact no extensive areas of superficial deposits occur in the district and they may be discounted as a major source of water supply: water from such deposits is in any case liable to become polluted.

Hastings Beds. The Tunbridge Wells Sand is an aquifer of only local importance, due mainly to its variable lithology and thickness. Boreholes up to 6 in (152 mm) in diameter commonly yield from 200 to 500 gallons per hour (g.p.h.), although 900 g.p.h. (for 16 ft (4·9 m) drawdown) is recorded from a 4-in (101-mm) reducing to 3-in (76-mm) bore at Fletching [444 246]. In 1909 a borehole 4 in (101 mm) in diameter at Tunbridge Wells [589 407] yielded 2500 g.p.h. but this had decreased to 1400 g.p.h. by 1947. Both of these boreholes continued for some distance into the underlying Wadhurst Clay. Small diameter bores drilled in the bottom of shafts have not generally succeeded in significantly increasing the original yield. At Newick [407 222] a 4-in (101-mm) bore at the bottom of a shaft 92 ft (28·0 m) deep and 6 ft (1·8 m) in diameter yielded 900 g.p.h. Large diameter shafts without bores have yielded up to 800 g.p.h., as at Wadhurst [656 457] where this yield is obtained from a shaft 5 ft (1·5 m) in diameter and 27 ft (8·2 m) deep. The highest yield recorded from a single borehole into the aquifer in this area is from one 18 in (457 mm) in diameter reducing to 12 in (305 mm) at Frant [632 356] which yielded 15 000 g.p.h. for a drawdown of 74 ft (22·6 m). Works at East Grinstead [396 394] consisting of two shafts connected by a heading yielded 18 000 g.p.h. in 1957, but silting reduced the depth of one shaft, and the yield in 1962 was 15 000 g.p.h.

In the Tunbridge Wells Sand rest water levels range from 56 to 346 ft (17·1–105·5 m) above O.D., with the highest levels occurring beneath the highest ground. A site at Frant [632 356], where water levels have been measured monthly since April 1957 by the Tunbridge Wells Corporation, has shown a maximum fluctuation of water level of 6 ft (1·8 m) between 185 and 179 ft (56·4 and 54·6 m) above O.D., the height above O.D. of the well top being 185·5 ft (56·5 m).

Chemical analyses of groundwater are available from three wells sunk into Tunbridge Wells Sand. Two of these indicate a soft water, the total hardness being 7 and 27 milligrammes per litre (mg/l) respectively, the latter analysis also showing 386 mg/l total dissolved solids and 33 mg/l chlorine (as chloride). The hardness of the third sample was 180 mg/l, but in this well 61 ft (18·6 m) of clay splits the sand into two parts. Analyses from wells which also entered the underlying Wadhurst Clay indicate, in general, a slightly hard to hard water, due probably to the contact of groundwater with sulphates in the clay formation. The ranges and average values of four constituents of such mixed waters are given below:

	Range (mg/l)	Average value (mg/l)
Total dissolved solids	150 to 257	209
Chlorine as chloride	17 to 106	44
Total hardness	93 to 283	127
Non-carbonate hardness	31 to 253	100

The Wadhurst Clay is primarily a clay formation, but discontinuous bands of limestone and sandstone occur locally and some water is obtained from these. As examples, a 4½-in (114-mm) borehole at Rotherfield [516 275] gave 300 g.p.h., a 6-in (152-mm) bore at Brenchley [658 411] 900 g.p.h. and an 8-in (203-mm) reducing to 6-in (152-mm) bore at Burwash [654 253] with a thickness of 26 ft (7·9 m) of Tunbridge Wells Sand overlying the clay, but lined out, gave 1000 g.p.h. during a 48-hour test. In general, however, consistent supplies of groundwater, other than for domestic needs, cannot be expected from this formation. In places spring flow issues from the sandstone bands, the latter being best developed towards the base of the formation. Springs rising at or close to the base of the Wadhurst Clay have been reported at a number of sites, three localities especially noted being at Mark Cross [585 311], Bestbeech Hill [614 311] and Wadhurst [638 316]. Many surface streams, especially in the Rother catchment, rise as springs at, or close to, the junction between the Wadhurst Clay and the Ashdown Beds. Although some of these spring flows may originate in sandstone bands in the Wadhurst Clay, much of the flow is probably derived from the underlying Ashdown Beds.

The only available chemical analysis of groundwater from Wadhurst Clay showed a very hard water (total hardness 429 mg/l) with an appreciable mineral content (total dissolved solids 843 mg/l).

J

TABLE 2

Representative analyses of groundwaters from Wealden Series strata

(expressed in milligrammes per litre)

	A	B	C	D
Location	Danehill	Rotherfield	Mayfield	Mayfield
National Grid Reference	[414 273]	[566 305]	[595 278]	[587 246]
Analyst*	Unknown	(1)	(2)	(3)
Date	? 1935	Unknown	25.10.51	23.3.50
Classification and thickness (ft) of strata†	A 100 (30·5 m)	spring rising in A	A 250 (76·2 m)	?WdC 50 (15·2 m) A 48 (14·6 m)
Total solid residue (dried at 180°C)	501	262	140	415
Chlorine as chlorides	32	12	21	27
Total hardness ‡	220	43	45	192
Non-carbonate hardness ‡	59	34	nil	43
Carbonate hardness ‡	161	9	45	149
Alkalinity ‡	—	43	80	—
pH	4·1	6·9	6·9	6·7
N_2 as nitrates	—	—	absent	0·1
N_2 as nitrites	—	—	absent	—
Total iron	—	2	6·6	2·6
Manganese	—	—	0·3	0·3

TABLE 2 (continued)

	E East Grinstead [393 384]	F East Grinstead [396 394]	G Tunbridge Wells [589 407]
Location	East Grinstead	East Grinstead	Tunbridge Wells
National Grid Reference	[393 384]	[396 394]	[589 407]
Analyst*	(4)	(5)	(6)
Date	8.2.12	13.12.34	20.7.38
Classification and thickness (ft) of strata†	GrC 6 (1·8 m) LTW 106 (32·3 m) WdC 235 (71·6 m) A 681 (207·6 m) ?Pb 9 (2·7 m)	LTW 89¾ (27·4 m) WdC 3½ (1·1 m)	LTW 94¼ (28·7 m) WdC 103½ (31·5 m)
Total solid residue (dried at 180°C)	350	220	—
Chlorine as chlorides ,,	24	20	—
Total hardness‡ ,,	4	105	210
Non-carbonate hardness ‡ ,,	4	75	196
Carbonate hardness ‡ ,,	nil	30	14
Alkalinity ‡ ,,	—	—	8·4
pH ,,	—	—	5·2
N_2 as nitrates ,,	absent	2·4	—
N_2 as nitrites ,,	absent	—	—
Total iron ,,	2·0	—	0·2
Manganese ,,	—	—	—

*Analyst: (1) Railway Executive, Southern Region, Ashford Works Chemical Laboratory. (2) Counties Public Health Laboratories, London, S.W.1. (3) Public Analyst's Laboratory, Lewes, Sussex. (4) Cecil H. Cribb, Esq. (5) Royal Institute of Public Health, London, W.C.1. (6) Permutit Company Limited, London, W.4.

†GrC = Grinstead Clay; LTW = Lower Tunbridge Wells Sand; WdC = Wadhurst Clay; A = Ashdown Beds; Pb = Purbeck Beds.

‡As calcium carbonate.

The Ashdown Beds constitute the principal aquifer of the area, 94 per cent of groundwater abstraction in the district in 1964 being derived from them (see above). The formation shows both lithological and thickness variations, with consequent variations in the yields from boreholes. Yields from small diameter (up to 6 in (152 mm)) boreholes range from nil to 3000 g.p.h., the latter having been recorded from a 6-in (152-mm) diameter bore at Buxted [493 247] where the aquifer is confined beneath the Wadhurst Clay. The normal range of yields from small diameter bores where the aquifer is not confined is from 150 to 500 g.p.h., although at Maresfield [432 301] a 6-in (152-mm) bore gave 1800 g.p.h. on test. Yields from boreholes of a given diameter are commonly rather higher under the clay cover than at outcrop because the upper part of the Ashdown Beds yields most freely in this area; drilling through the overlying Wadhurst Clay into the top of the Ashdown Beds thus ensures that the full thickness of the water-bearing strata is penetrated. Larger diameter boreholes situated in valleys have commonly proved successful, especially when sited close to fault zones. An example of this is an 18-in (457-mm) diameter borehole at Groombridge [528 364] which overflowed naturally at 2800 g.p.h. in 1930, and gave a pumped yield of 32 000 g.p.h. This site is in a valley, on the upfaulted edge of a horst of Ashdown Beds, faulted against Tunbridge Wells Sand. Other examples of high yields from large diameter bores in valleys include one of 6000 g.p.h. from an 18-in (457-mm) reducing to 12 in (305 mm) bore at Rother-field [567 302], a second of 10 000 g.p.h. from a 27-in (686-mm) reducing to 20 in (508 mm) bore at Mayfield [595 278] and a third, also at Mayfield [605 274], of 15 000 g.p.h. from a borehole 12 in (305 mm) reducing to 8 in (203 mm) in diameter. The largest recorded yield from the Ashdown Beds in this district is 40 000 g.p.h.; this is obtained from a 21-in (533-mm) diameter bore at the bottom of a shaft at Forest Row [427 334], also situated in a valley.

Some wells in the Ashdown Beds in this district show a diminution of yield with time. This may be due to silting causing a reduction in the effective depth of the borehole, and consequently reducing the thickness of aquifer contri-buting to the yield, or to an influx of fine material into that part of the aquifer immediately adjacent to the borehole. In the latter case, there will be an increase in head loss and reduction in the efficiency of the well: sand screens and gravel packs are used in some cases to overcome this difficulty. An example of decrease in yield is afforded by the well at Groombridge [528 364] referred to above. In 1930 it yielded 32 000 g.p.h. but this had declined to 23 000 g.p.h. in 1935, and by 1950 the combined maximum yield of this borehole and of a second, 15 in (381 mm) diameter and 100 yd (91 m) to the south (the latter yielding 19 000 g.p.h. when drilled in 1936), was only 26 000 g.p.h.

Many large diameter shafts in the Ashdown Beds have been failures, probably because they were sunk into the less permeable lower part of the formation or were situated in an area where clays and silts predominated over coarser deposits. In many cases where the original yields of shafts were poor, boreholes were drilled at the bottom, but in general little or no improvement in yield resulted, and yields of such works do not, in general, exceed a few hundred g.p.h. For example, at Tunbridge Wells [582 402] a shaft 117 ft (35·7 m) deep and provided with two headings, was deepened to a total depth of 906 ft (276·1 m) by a borehole 10 in (254 mm) reducing to 6 in (152 mm) in diameter. The com-bined yield, however, was only 700 g.p.h. for 5 hours per day.

Springs are fairly common at some levels in the Ashdown Beds, and locally produce flows adequate for public supply. For example, two springs in this formation at Rotherfield [544 304 and 547 313] yield 5000 g.p.h. and 7000 g.p.h. respectively. Rest water levels in the Ashdown Beds range from 113 to 572 ft (34·4–174·3 m) A.O.D., the highest levels occurring beneath the higher ground and the lowest in the valleys. The maximum recorded fluctuation of water level is 68 ft (20·7 m), from a well at East Grinstead [393 384], the height of the well top being 418 ft (127·4 m) A.O.D. Smaller fluctuations occur over the lower ground, for example 9 ft (2·7 m) in a well at Eridge [543 344] where the height above O.D. of the well top is 180 ft (54·9 m). The maximum water level in the two wells was 247 and 185 ft (75·3, 56·4 m) A.O.D. respectively, the latter being an overflow.

Ranges and average values of some chemical constituents of groundwaters from the Ashdown Beds at outcrop are given below:

	Range (mg/l)	Average value (mg/l)
Total dissolved solids	125 to 500	260
Chlorine as chlorides	12 to 32	21
Total hardness	43 to 220	85
Non-carbonate hardness	27 to 59	44

The analyses show considerable variation in total hardness, probably due to the lithological variation of the aquifer. Up to 11 mg/l of total iron are recorded, and groundwater from this formation is commonly reported as being ferruginous. Representative analyses from individual wells in the Ashdown Beds and other formations are given in Table 2.

An analysis made in 1956 is available for the Chalybeate Spring, Pantiles, Tunbridge Wells, this being the only medicinal spring still in use in the area. The constituents, in mg/l, are given below:

Calcium as Ca	27·0
Magnesium as MgO	9·4
Sodium as Na_2O	27·5
Potassium as K_2O	9·7
Chlorine as Cl	41·0
Carbon dioxide as CO_2	231·6
Sulphur as SO_3	53·8
Manganese as MnO	7·1
Iron as FeO	47·9

R.A.

REFERENCES

ABBOTT, G. 1909. Excursion to Eridge and Tunbridge Wells. *Proc. Geol. Assoc.*, **21**, 207–9.

—— and HERRIES, R. S. 1898. Excursion to Crowborough. *Proc. Geol. Assoc.*, **15**, 450–2.

ALLEN, P. 1938. Ashdown Sand–Wadhurst Clay junction. *Geol. Mag.*, **75**, 560.

—— 1941. A Wealden soil bed with *Equisetites lyelli* (Mantell). *Proc. Geol. Assoc.*, **52**, 362–74.

—— 1947. Notes on Wealden fossil soil-beds. *Proc. Geol. Assoc.*, **57**, 303–14.

—— 1949a. Wealden Petrology: The Top Ashdown Pebble Bed and the Top Ashdown Sandstone. *Q. Jnl geol. Soc. Lond.*, **104**, 257–321.

—— 1949b. Notes on Wealden bone-beds. *Proc. Geol. Assoc.*, **60**, 275–83.

—— 1954. Geology and geography of the London–North Sea Uplands in Wealden times. *Geol. Mag.*, **91**, 498–508.

—— 1955a. Age of the Wealden in North-Western Europe. *Geol. Mag.*, **92**, 265–81.

—— 1955b. Age of the Wealden in North-Western Europe. *Geol. Mag.*, **92**, 512.

—— 1959. The Wealden Environment: Anglo-Paris Basin. *Phil. Trans. R. Soc.*, **242**, 283–346.

—— 1960a. Geology of the Central Weald: a Study of the Hastings Beds. *Geol. Assoc. Centenary Guide*, No. 24.

—— 1960b. Strand-Line Pebbles in the mid-Hastings Beds and the Geology of the London Uplands. General Features. Jurassic Pebbles. *Proc. Geol. Assoc.*, **71**, 156–68.

—— 1961. Strand-Line Pebbles in the mid-Hastings Beds and the Geology of the London Uplands. Carboniferous Pebbles. *Proc. Geol. Assoc.*, **72**, 271–86.

—— 1962. The Hastings Beds Deltas: Recent Progress and Easter Field Meeting Report. *Proc. Geol. Assoc.*, **73**, 219–43.

—— 1965. L'âge du Purbecko-Wealdien d'Angleterre. *Mem. Bureau Recherches géol. min.*, No. 34 (*Colloque sur le Crétacé inférieur*), 321.

—— 1967a. Origin of the Hastings Facies in North-Western Europe. *Proc. Geol. Assoc.*, **78**, 27–105.

—— 1967b. Strand-Line Pebbles in the mid-Hastings Beds and the Geology of the London Uplands. Old Red Sandstone, New Red Sandstone and other Pebbles. Conclusion. *Proc. Geol. Assoc.*, **78**, 241–76.

ANDERSON, F. W. 1962. Correlation of the Upper Purbeck Beds of England with the German Wealden. *L'pool Manch'r geol. Jnl*, 3,(1), 21–32.

—— and BAZLEY, R. A. 1971. The Purbeck Beds of the Weald (England). *Bull. geol. Surv. Gt Brit.*, No. 34.

—— —— and SHEPHARD-THORN, E. R. 1967. The sedimentary and faunal sequence of the Wadhurst Clay (Wealden) in boreholes at Wadhurst Park, Sussex. *Bull. geol. Surv. Gt Brit.*, No. 27, 171–235.

ANON. 1961. Code of Stratigraphical Nomenclature; American Commission on Stratigraphic Nomenclature. *Bull. Amer. Assoc. Pet. Geol.*, **45**, No. 5, 654, article 14(a).

—— 1962. The Surface Water Year-Book of Great Britain, 1960–61. Min. Housing and Local Govt. H.M.S.O.

BARROW, G. 1915. Report of an Excursion to East Grinstead. *Proc. Geol. Assoc.*, **26**, 120–2.

BAZLEY, R. A. 1966. In *A. Rep. Inst. geol. Sci. for 1965*, Part I.

BAZLEY, R. A. and BRISTOW, C. R. 1965. Field Meeting to the Weald of East Sussex. *Proc. Geol. Assoc.*, **76**, 315–20.

BIRD, E. C. F. 1956. The Geomorphology of the Upper Medway basin. *S. East. Nat.*, **51**, 26–31.

—— 1958. Some aspects of the geomorphology of the High Weald. *Trans. Inst. Brit. Geor.*, **25**, 37–43.

—— 1964. Tor-like Sandrock Features in the Central Weald. *20th International Geographical Congress* (London, 1964), Abstract.

BOSWELL, P. G. H. 1918. *A Memoir on British Resources of Sands and Rocks used in Glass Making*. London.

BUCHAN, S. 1938. Notes on Some Outliers of Grinstead Clay around Tunbridge Wells, Kent. *Proc. Geol. Assoc.*, **49**, 407–9.

—— ROBBIE, J. A., HOLMES, S. C. A., EARP, J. R., BUNT, E. F. and MORRIS, L. S. O. 1940. Water Supply of South-East England from Underground Sources. *Geol. Surv. Wartime Pamphlet*, No. 10, Part VI.

CALLOMON, J. H. and COPE, J. C. W. 1971. The Stratigraphy and Ammonite succession of the Oxford and Kimmeridge clays in the Warlingham Borehole. *Bull. geol. Surv. Gt Brit.*, No. 36, 147–76.

CASEY, R. 1955a. The Neomiodontidae, a new family of the Arcticacea (Pelecypoda). *Proc. Malac. Soc. Lond.*, **31**, 208–22

—— 1955b. The pelecypod family Corbiculidae in the Mesozoic of Europe and the Near East. *Jnl Washington Acad. Sci.*, **45**, 366–72.

—— 1961. The Stratigraphical Palaeontology of the Lower Greensand. *Palaeontology*, **3**, 487–621.

—— 1963. The Dawn of the Cretaceous Period in Britain. *Bull. S.-E. Un. sci. Socs*, **117**, 1–15.

—— 1964. The Cretaceous period. *Q. Jnl geol. Soc. Lond.*, **120S**, 193–202.

CATTELL, C. S. 1970. Preliminary research findings relating to the bloomery period in the iron industry in the upper basin of the eastern Rother. *Bull. Hist. Metall. Grp*, **4**, No. 1, 18–20.

CHALONER, W. G. 1962. Rhaeto-Liassic plants from the Henfield Borehole. *Bull. geol. Surv. Gt Brit.*, No. 19, 16–28.

CLEERE, H. 1970. The Romano-British Industrial Site at Bardown, Sussex. *Sussex Archaeol. Soc. Occ. Pap.*, No. 1.

COLE, M. J., MATTHEWS, A. M. and ROBERTSON, A. S. 1965. Records of Wells in the Area of New Series One-Inch (Geological) Lewes (319), Hastings (320) and Dungeness (321) sheets. *Water Supply Pap. geol. Surv. Gt Brit., Well Cat. Ser.*

COOLING, C. M. and OTHERS. 1968. Record of Wells in the Area of New Series One-Inch (Geological) Reigate (286) and Sevenoaks (287) sheets. *Water Supply Pap. Inst. geol. Sci., Well Cat. Ser.*

DINES, H. G. and EDMUNDS, F. H. 1933. Geology of the Country around Reigate and Dorking. *Mem. geol. Surv. Gt Brit.*

—— BUCHAN, S., HOLMES, S. C. A. and BRISTOW, C. R. 1969. Geology of the Country around Sevenoaks and Tonbridge. *Mem. geol. Surv. Gt Brit.*

DREW, F. 1861. On the Succession of Beds in the "Hastings Sand" in the Northern Portion of the Wealden Area. *Q. Jnl geol. Soc. Lond.*, **17**, 271–86.

EDMUNDS, F. H. 1928. Wells and Springs of Sussex. *Mem. geol. Surv. Gt Brit.*

—— 1935. The Wealden District. *Brit. Reg. Geol.*

FALCON, N. L. and KENT, P. E. 1960. Geological Results of Petroleum Exploration in Britain 1945–1957. *Mem. geol. Soc. Lond.*, **2**.

FITTON, W. H. 1824. Inquiries respecting the Geological Relations of the Beds between the Chalk and Purbeck Limestone in the South East of England. *Ann. Phil.* (2), **8**, 365–485.

GALLOIS, R. W. 1963. In *Summ. Prog. geol. Surv. Gt Brit. for 1962*.

—— 1964. Field Meeting to the Haywards Heath Area, Sussex. *Proc. Geol. Assoc.*, **75**, 361–6.

—— 1965a. In *Summ. Prog. geol. Surv. Gt Brit. for 1964*.

—— 1965b. The Wealden District. *Brit. Reg. Geol.*

—— 1970. Written discussion to paper taken as read: 3 March 1967. *Proc. Geol. Assoc.*, **81**, 169–72.

GREGORY, J. W. 1908. *Report on the prospects of Iron Mining and Smelting in the neighbourhood of The Sheffield Park, Sussex*. Brighton.

HALL, S., MILNER, H. B. and SWEETING, G. S. 1933. A Traverse of the Central Weald, Tunbridge Wells, High Rocks and Eridge. *Proc. Geol. Assoc.*, **44**, 447–9.

HARVEY, B. I., MATTHEWS, A. M. and OTHERS. 1964. Records of Wells in the Area of New Series One-Inch (Geological) Tunbridge Wells (303) and Tenterden (304) sheets. *Water Supply Pap. geol. Surv. Gt Brit.*, *Well Cat. Ser.*

HERRIES, R. S. and ABBOTT, G. 1895. Excursion to Tunbridge Wells. *Proc. Geol. Assoc.*, **14**, 198–200.

HOLLINGWORTH, S. E., TAYLOR, J. H. and KELLAWAY, G. A. 1944. Large-Scale Superficial Structures in the Northampton Ironstone Field. *Q. Jnl geol. Soc. Lond.*, **100**, 1–44.

HOWITT, F. 1964. Stratigraphy and structure of the Purbeck inliers of Sussex (England). *Q. Jnl geol. Soc. Lond.*, **120**, 77–113.

HUGHES, N. F. 1958. Palaeontological evidence for the age of the English Wealden. *Geol. Mag.*, **95**, 41.

—— and MOODY-STUART, J. C. 1969. A method of stratigraphic correlation using early Cretaceous miospores. *Palaeontology*, **12**, 84–111.

KAYE, P. 1966. Lower Cretaceous Palaeontology of North-west Europe. *Geol. Mag.*, **103**, 257.

KELLAWAY, G. A. and TAYLOR, J. H. 1953. Early stages in the Physiographic Evolution of a portion of the East Midlands. *Q. Jnl geol Soc. Lond.*, **138** (for 1952), 343–66.

KENYON, G. H. 1967. *The Glass Industry of the Weald*. Leicester University Press.

LAKE, R. D. and THURRELL, R. G. (*in press*). The sedimentary and faunal sequence of the Wealden Beds in boreholes near Cuckfield, Sussex. *Rep. Inst. geol. Sci.*

LOBLEY, J. LOGAN. 1879. Excursion to Tunbridge Wells and Crowborough Beacon. *Proc. Geol. Assoc.*, **6–7**, 230–3.

LOCK, M. 1953. *Equisetites lyelli* (Mantell) at a new horizon in the Wadhurst Clay, near Pembury, Kent. *Proc. Geol. Assoc.*, **64**, 31–2.

MANTELL, G. A. 1822. *The Fossils of the South Downs*. London.

—— 1827. *Illustrations of the Geology of Sussex*. London.

MARGARY, I. D. 1948. *Roman Ways in the Weald*. London.

MICHAELIS, E. R. 1969. Geology of the Haywards Heath District. *Proc. Geol. Assoc.*, **79** (for 1968), 525–48.

MILNER, H. B. 1923a. Notes on the Geology and Structure of the Country around Tunbridge Wells. *Proc. Geol. Assoc.*, **34**, 47–55.

—— 1923b. The Geology of the Country around East Grinstead, Sussex. *Proc. Geol. Assoc.*, **34**, 283–300.

MILNER, H. B. 1924. The Geology of the country between Goudhurst (Kent) and Ticehurst (Sussex). *Proc. Geol. Assoc.*, **35,**, 383–94.

NORRIS, G. 1969. Miospores from the Purbeck Beds and marine Upper Jurassic of southern England. *Palaeontology*, **12**, 574–620.

POOLE, E. G., WILLIAMS, B. J. and HAINS, B. A. 1968. Geology of the Country around Market Harborough. *Mem. geol. Surv. Gt Brit.*

PRESTWICH, J. and MORRIS, J. 1846. On the Wealden Strata exposed by the Tunbridge Wells Railway. *Q. Jnl geol. Soc. Lond.*, **2**, 397–405.

REEVES, J. W. 1948. Surface Problems in the Search for Oil in Sussex. *Proc. Geol. Assoc.*, **59**, 234–69.

SCHUBERT, H. R. 1957. *History of the British Iron and Steel Industry from c. 450 B.C. to A.D. 1775.* London.

SHEPHARD-THORN, E. R. 1967. In *A. Rep. Inst. geol. Sci. for 1966.*

—— SMART, J. G. O., BISSON, G. and EDMONDS, E. A. 1966. Geology of the Country around Tenterden. *Mem. geol. Surv. Gt Brit.*

STRAKER, E. 1931. *Wealden Iron.* London.

—— and MARGARY, I. D. 1938. Ironworks and Communications in the Weald in Roman times. *Geogrl Jnl*, **92**, 55–60, and map facing p. 96.

SWEETING, G. S. 1944. Wealden Iron Ore and the History of its Industry. *Proc. Geol. Assoc.*, **55**, 1–20.

—— 1945. Field Meeting at Southborough, Tunbridge Wells, and Eridge. *Proc. Geol. Assoc.*, **56**, 153–5.

TAYLOR, J. H. 1963. Sedimentary features of an ancient deltaic complex: The Wealden Rocks of southeastern England. *Sedimentology*, **2**, 2–28.

—— 1951. Sedimentation problems of the Northampton Sand Ironstone, pp. 74–85; *in* Hallimond and others. Constitution and Origin of Sedimentary Iron Ores: A Symposium. *Proc. Yorks. geol. Soc.*, **28**, 61–101.

TERRIS, A. P. and BULLERWELL, W. 1965. Investigations into the underground structure of southern England. *Advmt Sci. Lond.*, **22** (98), 232–52.

THURRELL, R. G., WORSSAM, B. C. and EDMONDS, E. A. 1968. Geology of the Country around Haslemere. *Mem. geol. Surv. Gt Brit.*

TOPLEY, W. 1875. The Geology of the Weald. *Mem. geol. Surv. Gt Brit.*

WALTERS, R. C. S. 1962. *Dam Geology.* London.

WEEKS, A. G. 1969. The stability of natural slopes in South East England as affected by periglacial activity. *Q. Jnl Engng Geol.*, **2**, 49–61.

WHITAKER, W. 1908. Water Supply of Kent. *Mem. geol. Surv. Gt Brit.*

—— 1911. Water Supply of Sussex. *Mem. geol. Surv. Gt Brit.*

—— 1912. Water Supply of Surrey. *Mem. geol. Surv. Gt Brit.*

—— and REID, C. 1899. The Water Supply of Sussex from underground sources. *Mem. geol. Surv. Gt Brit.*

WOOLDRIDGE, S. W. and LINTON, D. L. 1955. *Structure, Surface and Drainage in South-east England.* London.

WORSSAM, B. C. 1963. Geology of the Country around Maidstone. *Mem. geol. Surv. Gt Brit.*

—— 1964. Iron Ore Workings in the Weald Clay of the Western Weald. *Proc. Geol. Assoc.*, **75**, 529–46.

—— 1964b. Written contribution to the discussion of a paper previously taken as read: 6 March 1964. *Proc. Geol. Assoc.*, **75**, 637–5.

—— and IVIMEY-COOK, H. C. 1971. The Stratigraphy of the Geological Survey Borehole at Warlingham, Surrey. *Bull. geol. Surv. Gt Brit.*, No. 36, 1–146.

Appendix I

SECTIONS OF BOREHOLES

The following two logs, for the Ashdown Nos. 1 and 2 boreholes, are condensed from the descriptions prepared for BP Petroleum Development Limited by their geologists and are published with the permission of that company. From an examination of the cores recovered some additional comments by officers of the Institute of Geological Sciences have been inserted. A revised stratigraphical account with notes on the fauna is given on pp. 16–24.

Ashdown No. 1 Borehole

Height above O.D. (Rotary Table) 623 ft (190 m), six-inch TQ 53 SW.

Site 1000 yd (914 m) W. of Crowborough Hospital. National Grid Ref. TQ 5006 3035.

Drilled in 1954–5 by BP Petroleum Development Limited (D'Arcy Exploration Company Limited).

Specimens from the cores were collected by R. V. Melville, R. Casey and M. J. Hughes.

Description of Strata	Thickness Ft	in	Depth Ft.	in
No samples	15	0	15	0
Wealden				
Rich yellowish brown sandstone and silt	19	0	34	0
Light brown sandy clay	41	0	75	0
Brown and grey silt, very fine-grained sandstone with lignite specks	30	0	105	0
Grey and brown sandy clay	10	0	115	0
Greyish brown silt	15	0	130	0
Grey sandy clay	10	0	140	0
Brown and grey lignitic silt	15	0	155	0
Brown and grey sandy clay	15	0	170	0
Grey, white and brown fine- to medium-grained sandstone, locally coarse grained	195	0	365	0
Grey and brown clay with silt bands	114	0	479	0
Purbeck Beds				
Upper Purbeck (141 ft)				
Grey and brown clay with silt bands	6	0	485	0
Fine-grained hard grey lignitic sandstone	11	0	496	0
Grey shale with thin bands of grey silt	69	0	565	0
Thin bands of white crystalline limestone	20	0	585	0
Grey micaceous and carbonaceous silt	35	0	620	0
Middle Purbeck (235 ft)				
Grey friable calcareous shale with thin (3-in) detrital shelly limestone, abundant bivalves	115	0	735	0
Banded hard grey calcareous silt, plant remains	10	0	745	0

128

Description of Strata	Thickness		Depth	
	Ft	in	Ft	in
Grey calcareous shale, thin shelly limestone bands, seat-earths and plant remains	40	0	785	0
Grey calcareous mudstones, bivalves and ostracods, iron-stone nodules (Cinder Beds)	11	7	796	7
Grey calcareous mudstone	8	5	805	0
Hard banded dark limestone, shelly at base	5	0	810	0
Grey calcareous mudstones, occasional thin limestones, greyish green 'seatearths' and black shales	45	0	855	0

Lower Purbeck (244 ft)

Greyish green shales, grey calcareous banded shales and silty bands, occasional thin limestones	80	0	933	0
Grey crystalline gypsum with shale partings	7	0	942	0
Greenish grey mudstone, locally laminated, mudcracks, slumping and contorted bedding around 1010–1030 ft ..	92	0	1034	0
Thin nodular algal limestone, dolomitic limestone and anhy-drite	26	0	1060	0
Greenish grey mudstone	8	0	1068	0
Massive anhydrite in two main beds split by mudstone, each with shale intercalations which contain anhydrite nodules	31	0	1099	0

Portland Beds (about 82 ft)

Dark grey silty shale and mudstone, locally with current bed-ding and ripple marks, occasional sandy laminae and with two argillaceous limestones near base. Passes down gradually into more shaly beds	82	0	1181	0

Kimmeridge Clay (about 1839 ft)

Medium and dark grey silty shales and calcareous shales. Shell beds recorded at 1200–20, 1270–90 ft and at about 1322, 1350, 1400, 1450, 1590, 1610 and 1690 ft	514	0	1695	0
Dark grey mudstone with bivalve fragments	2	0	1697	0
Dark grey hard argillaceous limestone	103	0	1800	0
Brownish bituminous shale with some dark grey calcareous shale; shell bed recorded at 2020 ft	283	0	2083	0
Dark grey blocky calcareous shales with molluscs	3	0	2086	0
Greyish black calcareous shales and thin grey argillaceous limestones, local bituminous shale, sandy calcareous shale near base	331	0	2417	0
Dark grey silty shale	10	0	2427	0
Grey silty argillaceous limestone and dark grey calcareous silty shale	43	0	2470	0
Dark grey calcareous mudstone, thin current-bedded muddy calcareous sandstone at about 2565 ft	175	0	2645	0
Dark grey silty mudstone with silt laminae, some cross bed-ding, micaceous, top slightly calcareous but more cal-careous below 2700 ft	298	0	2943	0
Shelly grey sandy limestone and pale grey sandstones ..	52	0	2995	0
Dark grey calcareous mudstones	25	0	3020	0

Corallian Beds (maximum about 384 ft)

Dark grey often friable shale, some pyritic and calcareous shales and three thin bands of light grey limestone ..	384	0	3404	0

Description of Strata	Thickness Ft in		Depth Ft in	

Oxford Clay (about 317 ft)
Dark grey calcareous shales and mudstone (about 100 ft) overlying dark brown bituminous, shelly, pyritic mudstone   317   0   3721   0

Kellaways Beds (about 31 ft)
Dark grey hard pyritic oolitic limestone   0   6   3721   6
Grey locally pyritic sandstone, some carbonaceous shaly laminae, bioturbated muddy sandstone, *Liostrea*   ..   30   6   3752   0

?Cornbrash (?11 ft)
Grey mudstone and grey limestone with scattered ooliths   ..   11   0   3763   0

Forest Marble (?38 ft)
Grey fine-grained limestone and oolitic limestone, echinoderm and shell fragments abundant, erosion surface at base   38   0   3801   0

Great Oolite (about 143 ft)
Fawn hard compact oolitic limestone with occasional partings of dark grey silty mudstone   143   0   3944   0

Fuller's Earth (about 90 ft)
Dark grey calcareous mudstone   44   0   3988   0
Grey fine-grained silty limestone, partings of calcareous mudstone   15   0   4003   0
Dark grey calcareous mudstone, slightly pyritic   31   0   4034   0

Inferior Oolite (about 411 ft)
Dark grey compact hard argillaceous and crystalline limestone   51   0   4085   0
Grey and greyish buff oolitic limestone with some softer bands   230   0   4315   0
Light grey and white, non-oolitic limestone   20   0   4335   0
Buff oolitic limestone and non-oolitic crystalline limestone   ..   110   0   4445   0

Lias (?Upper) (93 ft penetrated)
Reddish brown ferruginous limestone and ironstone   ..   27   0   4472   0
Green iron-rich siltstone   2   0   4474   0
Reddish brown oolitic limonitic ironstone   21   0   4495   0
Dark grey sandy limestone and fine-grained sandstone   ..   14   0   4509   0
Reddish brown oolitic limonitic ironstone   6   0   4515   0
Conglomerate, small reddish brown limonitic pebbles   ..   0   4   4515   4
Medium grey micaceous siltstone, calcareous siltstone and silty limestone   22   8   4538   0

Ashdown No. 2 Borehole

Height above O.D. (Rotary Table) 585·4 ft (178·42 m), six-inch TQ 52 NW.

Site 1200 yd (1·1 km) S. of Crowborough Hospital. National Grid Ref. TQ 5107 2924.

Drilled in 1955 by BP Petroleum Development Limited (D'Arcy Exploration Company Limited).

Specimens from the cores were collected by R. V. Melville.

Description of Strata	Thickness Ft in		Depth Ft in	
No samples	15	0	15	0

Description of Strata	Thickness Ft in	Depth Ft in

Wealden

White and pale brown fine-grained sandstone, nodules of clay ironstone throughout, pyrite and carbonaceous flecks with occasional larger pieces .. .. 273 0 — 288 0

Dark greyish brown siltstone and white fine-grained sandstone 60 0 — 348 0

White, and some pale brown, fine-grained sandstone with clay ironstone nodules .. .. 137 0 — 485 0

Soft greyish brown and greenish calcareous clayey siltstone and clay. A little white, calcareous fine-grained sandstone near the top .. .. 32 0 — 517 0

Soft greenish grey and brownish grey calcareous clay and silty clay. Carbonaceous flecks .. .. 42 0 — 559 0

Purbeck Beds

Upper Purbeck (141 ft)

Soft greenish grey and brownish grey calcareous clay and silty clay, clay ironstone nodules, near the top, carbonaceous flecks throughout .. .. 81 0 — 640 0

Soft grey and brownish grey shales with thin fine-grained sandstones locally calcareous and sideritic .. .. 60 0 — 700 0

Middle Purbeck (235 ft)

Soft greyish brown shale and silty shale .. .. .. 40 0 — 740 0

Dark grey shales with thin bands of shell-detrital limestone. Some silty shale .. .. 72 0 — 812 0

Hard grey to pale brown fine-grained sandstone passing down into siltstone .. .. 26 0 — 838 0

Dark grey and greyish black shales locally calcareous and shelly throughout .. .. 27 0 — 865 0

Dark grey shales, locally calcareous, shelly throughout (Cinder Beds) .. .. 12 6 — 877 6

Hard greyish brown fine-grained silty limestone .. .. 7 6 — 885 0

Dark grey calcareous shale and mudstone with a few bivalves and gastropods .. .. 50 0 — 935 0

Lower Purbeck (245 ft)

Dark grey calcareous shale and mudstone with a few bivalves and gastropods .. .. 45 0 — 980 0

Hard greyish brown fine-grained silty limestone .. .. 2 0 — 982 0

Hard finely banded dark and pale grey calcareous mudstone, more calcareous towards base .. .. 40 0 — 1022 0

Granular intergrowth of dolomitic limestone, anhydrite and gypsum, bands of grey calcareous mudstone, black shale and fibrous gypsum .. .. 4 7 — 1026 7

Finely laminated dark and pale grey calcareous mudstone, thin bands of brown algal limestone at the top .. 85 5 — 1112 0

Hard dark greyish brown granular limestone .. .. .. 2 0 — 1114 0

Beds of massive white anhydrite and patches of gypsum, alternating with dolomitic limestone with dark grey calcareous mudstone containing some gypsum and anhydrite .. .. 66 0 — 1180 0

Portland Beds (about 70 ft)

Dark grey poorly bedded calcareous muddy siltstone and silty mudstone, pyritic near top; local cross bedding and burrowed beds; a 6-in limestone at 1231 ft; scattered fish remains and bivalves. Passes down into .. .. .. 70 0 — 1250 0

Description of Strata	Thickness		Depth	
	Ft	in	Ft	in
Kimmeridge Clay (about 1730 ft)				
Greenish grey and dark grey calcareous shale, a few shells and pyrite fragments	100	0	1350	0
Dark grey calcareous shale with a few shells and pyrite fragments	352	0	1702	0
Hard brown fine-grained muddy limestone	3	0	1705	0
Grey and dark grey calcareous shale and mudstone	48	0	1753	0
Hard pale grey fine-grained muddy limestone	69	0	1822	0
Hard dark grey calcareous mudstone and thin muddy limestones	20	0	1842	0
Hard dark grey calcareous mudstone and shale; more silty horizons occur at 2270–303 ft, 2320–87 ft, 2560–5 ft, 2635–55 ft, 2705–800 ft and brownish grey mudstone and muddy limestone from 2303–20 ft	1068	0	2910	0
Hard grey oolitic shelly sandy limestone	2	0	2912	0
Dark grey calcareous sandy mudstone, sandstone and calcareous mudstone	68	0	2980	0
Corallian Beds (about 368 ft)				
Hard grey calcareous mudstone and muddy limestone ..	10	0	2990	0
Dark grey calcareous mudstone and shale with some thin limestone bands	170	0	3160	0
Dark grey non-calcareous shale	77	0	3237	0
Dark grey calcareous mudstone, pale grey oolitic limestone band from 3344 ft	111	0	3348	0
Oxford Clay (about 304 ft)				
Dark grey and brownish grey pyritic shale and mudstone, locally slightly silty	187	0	3535	0
Dark brownish grey calcareous silty and sandy mudstone ..	20	0	3555	0
Dark brownish grey calcareous slightly silty mudstone ..	95	0	3650	0
Hard dark grey silty mudstone	2	0	3652	0
Kellaways Beds (about 30 ft)				
Massive pale and dark grey muddy fine-grained sandstone, locally calcareous and shelly	30	0	3682	0
?Cornbrash (?11 ft)				
Dark grey pyritic muddy siltstone, shelly and calcareous at the base	11	0	3693	0
Forest Marble (22 ft)				
Hard massive greyish brown oolitic limestone, muddy and dark grey near base, thin band of highly calcareous fine-grained sandstone below 3713 ft	22	0	3715	0
Great Oolite (158 ft)				
Hard massive greyish brown oolitic limestone, muddy and dark grey near base	158	0	3873	0
Fuller's Earth (84 ft)				
Dark grey calcareous mudstone, slightly silty in places, bands of muddy limestone at 3905–25 ft	84	0	3957	0
Inferior Oolite (about 379 ft)				
Hard pale grey and buff oolitic limestone with some dark greyish brown fine-grained limestone below 4200 ft ..	379	0	4336	0

Description of Strata	Thickness		Depth	
	Ft	in	Ft	in
Lias (about 1292 ft)				
Dark purplish red calcareous oolitic ironstone, a thin bed of dark greenish grey calcareous siltstone at 4348 ft	42	0	4378	0
Grey and dark grey calcareous siltstone, fine-grained sandstones	93	0	4471	0
Dark grey micaceous slightly calcareous mudstone locally slightly silty, greyish brown muddy limestone at 4563–6 ft	179	0	4650	0
Dark grey micaceous slightly calcarous silty mudstone and siltstone and pale fine-grained limestone	22	0	4672	0
Hard grey and greyish green sandy limestone with some chamosite	39	0	4711	0
Grey to dark grey micaceous silty mudstone, often calcareous and with a few thin bands of limestone and fine sandstone	103	0	4814	0
Dark grey micaceous silty mudstones, locally calcareous with thin greyish brown calcareous mudstone and muddy limestone	314	0	5128	0
Alternation of dark grey micaceous calcareous silty mudstone and grey and brown muddy limestone. Some beds of sandy limestone	142	0	5270	0
Dark grey micaceous pyritic silty mudstone with a few bands of grey silty limestone	60	0	5330	0
Grey silty and some sandy limestones alternating with dark grey slightly silty mudstone	60	0	5390	0
Dark grey muddy silty limestone, thin bands of dark grey mudstone	105	0	5495	0
Grey calcareous fine-grained sandstone, sandy limestone and dark grey mudstone	35	0	5530	0
Grey muddy silty and shelly limestones with thin bands of mudstone	60	0	5590	0
Hard dark grey muddy limestone, silty, sandy limestone and with beds of pale dolomitic limestone, pale shale and fine-grained sandstone	38	0	5628	0
Beds of Unknown Age (81 ft)				
Red mudstone	2	0	5630	0
Brecciated dark crimson calcareous mudstone	22	0	5652	0
Dark purplish and brownish red silty and calcareous mudstones; a dark brownish red fine-grained sandstone (7 ft), top at 5683 ft	57	0	5709	0

Completed at 5709 ft

Tangier Farm (Frant) No. 1

Height above O.D. 373·2 ft (113·75 m).
Sited 280 yd (256 m) at 160° from the Waterworks [5876 3448].
Drilled in 1962.

Description of Strata	Thickness		Depth	
	Ft	in	Ft	in
Wealden				
Ashdown Beds				
Core lost	11	6	11	6
White, very fine-grained, cross-bedded, permeable sandstone. Dendritic iron staining; some carbonaceous plant debris..	4	0	15	6

Description of Strata	Thickness		Depth	
	Ft	in	Ft	in
Core lost	7	6	23	0
White, very fine-grained sandstone, as above	0	9	23	9
Grey silt with plant debris, passing down into	0	1	23	10
Very fine-grained, cross-bedded, silty iron-stained sandstone	0	3	24	1
Compact, pale to mid-grey silt with some very fine sand; abundant plant debris	0	8	24	9
Finely interlaminated grey shales and thin siltstones	0	4	25	1
Core lost	7	11	33	0
Grey, iron-stained, cross-bedded silt and fine sand with plant debris; bivalves (*Unio*) at 35 ft 6 in	2	8	35	8
Core lost	7	4	43	0
Grey, iron-stained, medium-grained, cross-bedded sandstone with plant debris	0	2	43	2
Pale grey, fine silty sandstone with plant debris	1	6	44	8
Dark grey, interlaminated mudstone and silt with plant debris; laminae disturbed by 'bioturbation'	0	6	45	2
Core lost	7	10	53	0
Fine-grained, silty, cross-bedded sandstone	1	0	54	0
Core lost	29	0	83	0
Compact, pale grey, cross-bedded silt with thin streaks of shale, much plant debris and rare fish teeth; vertical roots in lower 6 in (0·25 in diameter)	4	9	87	9
Core lost	5	3	93	0
Compact, laminated, pale grey silt with plant debris	2	4	95	4
Olive-green mudstone with interlaminated thin siltstones	1	8	97	0
Cross-laminated siltstones with shale partings; rootlets and plant debris	0	7	97	7
Fine, white, cross-bedded sandstone with plant debris	0	5	98	0
Core lost	2	0	100	0
Fine, silty sandstone, pale grey but darker and more silty in central 1ft; micaceous, and full of plant debris	3	6	103	6
Very fine, white sandstone with some silts; abundant plant debris	2	6	106	0
Sandstone as above interlaminated with larger proportion of silts; signs of bioturbation	4	0	110	0
Core lost	3	0	113	0
Olive-green, silty shale with traces of vertical roots	0	10	113	10
Hard, dark grey siltstone with *Equisetites in situ* at the top (vertical stems up to 0·4 in diameter)	0	10	114	8
Compact, mixed grey silt and mudstone; laminations disturbed	5	3	119	11
Compact, very fine, white sandstone with some silt	0	5	120	4
Mixed, compact grey silt and fine sand; affected by 'bioturbation'	2	0	122	4
Core lost	0	8	123	0
Interlaminated silt and mudstone mixed with some fine compact sandstone	4	0	127	0
Fine, white cross-bedded sandstone, coarser at top, siltier at base	0	8	127	8
Core lost	0	4	128	0
Fine, white sandstone with plant debris	0	9	128	9
Interlaminated silt and fine sandstone with plant debris, bioturbation evident	1	0	129	9

Description of Strata	Thickness		Depth	
	Ft	in	Ft	in
Fine-grained, cross-laminated sandstone with silty partings bearing plant debris	4	9	134	6
Compact grey silt with plant debris	4	0	138	6
Core lost	0	6	139	0
Grey silt with some darker mudstone containing plant debris	3	0	142	0
Core lost	1	0	143	0
Mixed cross-bedded silt and mudstone	1	0	144	0
Interlaminated fine sand and grey silt affected by bioturbation	0	10	144	10
Core lost	0	2	145	0
Grey silt with plant debris and a few bivalves, bioturbation	6	2	151	2
Greyish green silty mudstone	0	2	151	4
Hard, fine, white, rather silty cross-laminated sandstone ..	0	7	151	11
Mixed, very fine sandstone and grey silt with plant debris, becomes more silty downwards	3	1	155	0
Compact, laminated grey silt with plant debris	9	0	164	0
Core lost	1	6	165	6
Very fine, cross-laminated sandstone with some silt.. ..	1	10	167	4
Cross-laminated, grey silt	1	0	168	4
Dark grey silty mudstone packed with carbonaceous plant fragments in upper part; occasional plant fragments in lower part	1	5	169	9
Grey silt	0	1	169	10
Hard, white, cross-bedded, fairly porous silty sandstone ..	8	2	178	0
Very fine-grained, white, cross-laminated sandstone with some silt	1	9	179	9
Dark grey silt with some fine sand	0	10	180	7
Very fine, white, cross-laminated sandstone with some silt..	2	2	182	9
Dark grey silt with carbonaceous plant debris	1	0	183	9
Core lost	5	3	189	0
Hard, laminated, pale grey silt with plant debris	1	8	190	8
Dark grey, sandy mudstone, full of carbonaceous roots (cf. seatearth)	1	0	191	8
Hard, white, fine-grained sandstone with plant fragments; becoming silty towards base with dark silty laminations	4	0	195	8
Dark grey, banded silt	0	9	196	5
Olive-green mudstone	0	6	196	11
Mudstone and silty mudstone with fine sandstone generally olive-green in colour resting with sharp, but irregular junction on:	1	3	198	2
Hard white, fine-grained sandstone with plant fragments ..	0	5	198	7
Core lost	1	5	200	0
Dark grey muddy siltstone with sphaerosiderite concretions (ironshot)	0	2	200	2
Dark compact siltstone with plant fragments	0	4	200	6
Mixed, laminated fine sand and silt, affected by 'bioturbation'	0	4	200	10
Hard, white, fine-grained, ironshot sandstone, porous, micaceous; basal 6 in rich in plant material	2	0	202	10
Mixed, banded sandstone/silt, rock darkens rapidly passing down into compact dark grey banded silt	1	10	204	8
Core lost	1	4	206	0
Dark grey silt with fine sand laminae, plant debris common	2	8	208	8
As above but more sandy rock with plant debris	0	10	209	6
Dark grey silt, laminated in upper part	0	11	210	5

K

Description of Strata	Thickness Ft	in	Depth Ft	in
Greyish green sandy silt with pellets of mudstone (? bone bed) up to	0	0½	210	5½
Core lost	1	3½	211	9
Hard, very fine-grained, cross-laminated silty sandstone ..	0	3	212	0
Core lost	1	0	213	0
Dark grey, silty mudstone, faint greyish green mottling ..	2	0	215	0
Pale grey, cross-laminated silt with plant debris; passing down into:	0	11	215	11
Varied dark to medium grey silty mudstone shot with sphaerosiderite	0	11	216	10
Dark greyish green, silty mudstone	0	3	217	1
Uniform pale grey silt; passing down into:	1	0	218	1
Dark, compact, silty mudstone	0	10	218	11
Abrupt colour change to pale grey very fine-grained silty sandstone becoming less silty downward; passing down into	0	8	219	7
Very fine-grained, white sandstone, ironshot and not very porous	1	6	221	1
Compact, laminated, dark grey silt; laminations broken by bioturbation	1	11	223	0
Hard, very fine, cross-laminated sandstone; silty, in part, with abundant plant debris	2	0	225	0
Hard, compact, olive-green silt	0	10	225	10
Hard, fine-grained, silty sandstone especially silty in top few inches; micaceous	4	0	229	10
Core lost	2	0	231	10
Compact, mixed, cross-laminated fine sand and dark grey silt. Sharp junction with underlying	1	3	233	1
Dark grey shale	1	6	234	7
Core lost	1	5	236	0
Dark grey, silty mudstone with patchy dark red coloration, paler and more silty in lower part	5	3	241	3
Pale grey silty mudstone	0	11	242	2
Compact pale grey muddy silt	0	7	242	9
Very fine, pale grey silty sandstone having erosional contact with underlying	1	2	243	11
Compact, mid-grey silty mudstone with a few sphaerosiderite grains	1	1	245	0
Mid-grey silty mudstone with coarser, more abundant grains of sphaerosiderite and patchy red staining..	1	6	246	6
Compact grey silt	0	6	247	0
Core lost	1	0	248	0
Compact, greyish green silt with patchy red coloration and grains of sphaerosiderite	3	3	251	3
Hard, very fine-grained silty sandstone, not very porous. Patchy greyish green and red colouring	2	3	253	6
Compact greyish green and red silt	0	8	254	2
Mottled, compact, greyish green and red, muddy silt ..	4	8	258	10
Compact, mixed grey silt and very fine-grained sandstone ..	1	10	260	8
Dark grey silty mudstone with abundant plant debris; very micaceous; fragments of pyritized wood	2	6	263	2
Core lost	0	10	264	0
Dark grey silty mudstone	1	8	265	8

Description of Strata	Thickness Ft	in	Depth Ft	in
Hard grey silt with very fine sandstone; scattered ferruginous pellets, passing down into	0	6	266	2
Compact muddy dark grey siltstone with sphaerosiderite grains	2	9	268	11
Core lost	1	1	270	0
Compact muddy silt with sphaerosiderite grains	1	8	271	8
Dark, greyish green and red silty mudstone; vertical roots at top	6	6	278	2
Dark grey silty mudstone with plant debris	4	0	282	2
Mottled, grey and dark red silt, very compact throughout, sandy at base	3	0	285	2
Core lost	0	10	286	0
Mottled, soft, red and greyish green, muddy silt	1	3	287	3
Grey and red mottled silty mudstone	1	0	288	3
Grey shale	0	3	288	6
Compact, mottled, dark grey and red silty mudstone with some sphaerosiderite grains	3	6	292	0
Mottled siltstone crowded with minute sphaerosiderite grains	0	5	292	5
Mottled red and greyish green muddy silt with plant debris	0	10	293	3
Mottled red and greyish green silty mudstone	2	9	296	0
Core lost	1	0	297	0
Mottled silty mudstone	0	6	297	6
Mottled grey and red muddy siltstone	3	6	301	0
Compact, greenish grey silt with slight, patchy red staining	2	2	303	2
Compact, dark grey and red muddy siltstone	1	5	304	7
Silty mudstone, mottled	2	6	307	1
Core lost	0	11	308	0
Mottled, dark grey and red, silty mudstone	0	6	308	6
Compact, mottled red and green muddy silt	4	0	312	6
Core lost	0	6	313	0
Heavily red-stained compact muddy silt	3	3	316	3
Dark grey, compact silt..	2	2	318	5
Dark, greyish green and mottled muddy silt	4	4	322	9
Friable, mottled silty mudstone	0	4	323	1
Core lost	1	11	325	0
Mottled silty mudstone	1	5	326	5
Compact, dark grey muddy silt	1	0	327	5
Compact very hard, dark grey silt with red mottling ..	2	2	329	7
Mottled silty mudstone	1	6	331	1
Compact, mottled, grey and red silt	1	7	332	8
Core lost	1	4	334	0
Very hard, greyish green and red silt and very fine sandstone	2	0	336	0
Very friable, mottled greyish green and red silty mudstone..	3	2	339	2
Grey and red silt	1	0	340	2
Compact, greyish green silt, muddy and red stained at base	2	4	342	6
Friable red and greyish green silty mudstone	2	0	344	6
Red and greyish green muddy silt	2	1	346	7
Dark grey, red stained muddy siltstone with sphaerosiderite grains	0	8	347	3
Similar siltstone without grains	0	8	347	11
Very hard, pale grey, fine sandy silt	0	3	348	2
Finely interlaminated greyish green silt and mudstone with sphaerosiderite grains	1	10	350	0

Description of Strata	Thickness		Depth	
	Ft	in	Ft	in
Compact, red and grey silty mudstone	3	9	353	9
Hard grey and red muddy siltstone	1	3	355	0
Core lost	3	0	358	0
Compact, dark greyish green and red, very fine-grained silty sandstone, sparsely ironshot	0	7	358	7
Compact grey and red silt, becoming darker and more muddy downwards	5	0	363	7
Very hard, fine-grained, greyish green silty sandstone ..	0	9	364	4
Hard, grey and red mottled, micaceous silt	1	3	365	7
Hard, grey, very fine-grained silty sandstone..	0	3	365	10
Hard, cross-bedded and ?slumped grey silt and fine sand with banded appearance, abundant plant debris	5	6	371	4
'Mudstone conglomerate' containing angular pale grey mudstone fragments set in a yellowish matrix of coarse sand and mud with plant fragments	0	2	371	6
Compact, hard, dark grey silt with sphaerosiderite grains in part	1	6	373	0
Greyish green, red mottled, ironshot muddy silt	2	1	375	1
Core lost	7	11	383	0

E.R.S.-T., R.W.G.

Tangier Farm (Frant) No. 2

Height above O.D. 326·00 ft (99·36 m).
Sited 600 yd (549 m) at 103° from the Waterworks [5929 3659]. Drilled in 1962.

Description of Strata	Thickness		Depth	
	Ft	in	Ft	in

Wealden
Ashdown Beds

Core lost	17	0	17	0
Iron-stained cream coloured silty sandstone	0	4	17	4
Compact, banded pale grey muddy silt with fragmentary carbonaceous plant remains and poor bivalve fragments..	1	3	18	7
Olive-grey shale, mudstone and silty mudstone	0	5	19	0
Friable core lost; ? lithology as above	4	0	23	0
Hard compact interlaminated pale grey silt and dark grey muddy silt with poorly preserved Unionid at 24 ft 9 in ..	3	1	26	1
Dark and light grey silt with disturbed laminations; ripple mark at 26 ft 6 in with medium-grained sand in ripple crest, 0 to ¼ in thick	0	6	26	7
Finely micaceous interlaminated silt and sandy silt with lenses of coarser sand along ripple mark crests, 0 to ¼ in thick, at 27 ft 1 in and 27 ft 3 in	1	0	27	7
Hard, compact, interlaminated micaceous coarse pale grey silt and very silty sandstone to 28 ft 1 in passing down into finely laminated grey silt with fragmentary plant remains	1	3	28	10
Core lost	3	2	32	0
Laminated compact silt with ripple mark of fine white sand at 32 ft 6 in	1	0	33	0
Evenly laminated pale and dark grey compact silt with fragmentary plant remains; thin sandy 'biscuits' and ripple marks; shallow channel infilled with medium- to coarse-grained sand, 0 to ¼ in thick, at 37 ft 2 in	4	3	37	3

Description of Strata	Thickness		Depth	
	Ft	in	Ft	in
As above but more sandy	0	3	37	6
Core lost	5	6	43	0
Laminated dark and pale grey silt	0	3½	43	3½
Coarse sandstone with lignitic plant fragments; no pebbles present but equivalent of a pebble bed	0	1	43	4½
Laminated dark grey muddy silt	0	2½	43	7
Friable dark grey shale, mudstone and silty mudstone ..	1	5	45	0
Core lost	0	3	45	3
Hard compact porous brown and grey fine-grained sandstone; micaceous; poor plant fragments; patchy brown and greyish green iron staining throughout; more silty in basal 3 in	4	6	49	9
Laminated pale grey silt	0	9	50	6
Core lost	2	6	53	0
Evenly laminated grey silt with thin 'biscuits', lenses and ripples of fine silty sand; poor carbonaceous plant fragments; disturbed lamination in part	8	6	61	6
Coarse-grained sandstone; thin lenses of coarse sand occur in the silts for ½ in above and below this horizon.. ..	0	1	61	7
Laminated silt with thin lenses of fine silty sand	0	11	62	6
Core lost	0	6	63	0
Laminated pale grey silt with sandy lenses passing down into	0	6	63	6
More uniform pale grey silt with poor fragmentary plant remains	3	2	66	8
Dark olive-grey shale, mudstone and silty mudstone; core very friable and broken with slickensiding..	1	4	68	0
Laminated dark grey muddy silt and pale grey silt passing down into more uniform pale grey silt with a very poor bivalve impression; becoming dark and more muddy towards base	2	0	70	0
Dark olive-grey shale and mudstone; more silty at base ..	0	9	70	9
Laminated pale grey silt with carbonaceous plant fragments	1	3	72	0
Core lost	1	0	73	0
Laminated silt with poorly preserved bivalve at 73 ft 6 in, infilled worm tube at 73 ft 9 in	1	6	74	6
Poorly laminated compact muddy dark grey silt with a thin bed of coarse sand at 78 ft 9 in	8	3	82	9
Olive-grey friable shale and mudstone with abundant small rootlets	0	3	83	0
Laminated pale and dark grey silt with laminations disturbed by fine rootlets in growth position (bioturbation) ..	1	4	84	4
Laminated silts: sandy in part, micaceous, some bioturbation	3	0	87	4
Lithology as above but with 'rolled' and disturbed laminations due to ? slumping	0	8	88	0
Dark olive-grey shale and mudstone with thin paler silt laminations. Poorly preserved vertical stem, 0·4 in diameter, in growth position at 88 ft 4 in	1	0	89	0
Interlaminated silt and mudstone	0	2	89	2
Interlaminated fine-grained silty sandstone and darker grey silt; micaceous, poor plant remains; bioturbation in part, predominantly sandy	1	2	90	4
Fine-grained silty white sandstone with poor plant remains	0	10	91	2
Bioturbated laminae of silt and fine sand, predominantly silty	0	6	91	8

Description of Strata	Thickness		Depth	
	Ft	in	Ft	in
Core lost	1	4	93	0
Bioturbated laminae of fine silty sand and silt with poorly preserved fine rootlets at 94 ft 2 in and coarse sand, $\frac{1}{32}$ in thick, at 94 ft 6 in	2	4	95	4
Interlaminated fine sand and silt with rolled laminae, ?slumped	0	8	96	0
Hard, compact, silty, very fine-grained sandstone with disturbed lenses of interlaminated silt; micaceous with poor plant fragments and minute sphaerosiderite pellets randomly distributed throughout basal 2 in	3	3	99	3
Hard fine-grained sandstone with appreciable amounts of dark bioturbated silt increasing downwards and passing down into	2	6	101	9
Laminated bioturbated silt	0	8	102	5
Core lost	0	7	103	0
Interlaminated pale and dark grey silt with disturbed laminations	1	2	104	2
Olive-grey friable shale and mudstone with poorly preserved vertical stem, 0·4 in diameter, in position of growth at 105 ft and horizontal rhizomes, 0·2 in diameter, in positions of growth at 105 ft 9 in	2	8	106	10
Core lost	2	6	109	4
Laminated pale and dark grey silt, more muddy in upper 6 in, poor plant remains, hard and compact; bioturbated in part	3	8	113	0
Evenly laminated silt	0	8	113	8
Hard fine-grained silty pale grey sandstone with thin intercalations of darker, muddy silt; micaceous	0	9	114	5
As above but with strong bioturbation and becoming more silty downwards until predominantly silty and less micaceous	6	6	120	11
Hard compact pale grey silt with occasional thin muddy intercalations	0	10	121	9
Very dark grey silt crowded with carbonaceous plant fragments and with thin slickensided shale lenses, passing down into	1	3	123	0
More sandy dark grey silt with less carbonaceous plant material (although still abundant)	1	0	124	0
Dark grey silty fine-grained sandstone passing down into coarser and paler sandstone with abundant lignitic material	2	0	126	0
Dark silty sandstone, without carbonaceous material	1	0	127	0
Pale grey to white fine-grained sandstone; micaceous; plant fragments; silty and laminated in lower part and passing down into	1	6	128	6
Laminated pale grey silt	1	6	130	0
Core lost	3	0	133	0
Poorly laminated pale grey silt: Unionid at 134 ft. Darker and more muddy from 137 ft 6 in to 138 ft 6 in; fine rootlets from 139 ft to 139 ft 6 in	6	9	139	9
Laminated silty fine-grained sandstone with darker silt intercalations	0	4	140	1
Laminated silt with sporadic fine rootlets at 140 ft 6 in; ? slumping from 142 ft 6 in to 143 ft	3	11	144	0
Olive-grey friable mudstone	0	4	144	4

Description of Strata	Thickness Ft	in	Depth Ft	in
Pale grey very silty fine-grained sandstone; micaceous; poor plant fragments; passing down into laminated sand-silt ..	1	6	145	10
Laminated dark and light grey silt with some bioturbation; slightly micaceous; poorly preserved fine rootlets at 150 ft	6	2	152	0
Core lost	1	0	153	0
Laminated dark grey silt and pale sandstone, passing down into	1	3	154	3
Laminations in which sandstone predominates	0	9	155	0
Laminations darker and more silty	0	9	155	9
Laminations predominantly sandy with poorly preserved fine rootlets	0	9	156	6
Laminae intimately mixed, sand and silt	2	0	158	6
Core lost	4	6	163	0
Dark grey silt with thin sandy 'biscuits' riddled with lignitic plant debris	1	0	164	0
Olive-grey shale and silty mudstone; very friable and internally slickensided	0	3	164	3
Compact dark grey very muddy silt and silty mudstone; little lamination; well-preserved plant fragments	4	6	168	9
Laminated pale and dark grey silt	0	4	169	1
Interlaminated sand and silt with much included lignitic material	1	2	170	3
Compact, hard, olive-grey muddy silt..	0	6	170	9
Interlaminated silt and fine sand	1	3	172	0
Core lost	1	0	173	0
Laminated light and dark grey silt with some bioturbation	1	6	174	6
Hard, fine-grained, silty, micaceous, pale brown sandstone	1	0	175	6
Sandstone with interlaminated dark silty and carbonaceous material; fine bedding but very disturbed	1	2	176	8
Hard, compact, very dark grey, carbonaceous silt ..	0	3	176	11
Hard cream and white, compact, fine-grained silty ironshot sandstone with minor silty intercalations	2	4	179	3
Very friable silty mudstone; dark grey weathering to silvery grey	0	5	179	8
Core lost	1	0	180	8
Hard compact pale grey silty sandstone; fine-grained; thin dark clayey streaks; minute brown sphaerosideritic pellets at 181 ft 4 in, abundant for ¼ in and more sporadic for the next 3 in downwards	2	4	183	0
Hard, compact, fine-grained silty sandstone with sporadic plant debris	2	4	185	4
Silty sandstone with thin intercalations of silt and carbonaceous muddy silt	0	6	185	10
Interlaminated dark silt and pale silty sandstone	0	4	186	2
Dark grey muddy silt, finely micaceous, becoming muddier downwards, dermal denticle at 186 ft 11 in	0	10	187	0
Friable olive-grey shale and mudstone; fine rootlets at 187 ft 4 in; numerous small sphaerosideritic pellets from 187 ft 6 in to 187 ft 8 in	0	10	187	10
Yellow and greyish green silty mudstone rapidly passing down into	0	9	188	7
Muddy siltstone and light grey silt, sporadically ironshot, coarsening downwards and becoming paler in colour ..	1	9	190	4

Description of Strata	Thickness Ft in		Depth Ft in	
Core lost (probably in shale portion of above as rest of core is fairly continuous)	2	8	193	0
Hard, clean, porous, white, fine-grained sandstone; ironshot (sphaerosiderite) and micaceous, poor plant remains ..	1	2	194	2
Sandstone as above with muddy and carbonaceous intercalations	1	0	195	2
Compact, dark grey, muddy silt with plant fragments ..	1	4	196	6
Hard, compact, laminated pale and dark grey silt with bioturbation	0	11	197	5
Dark grey compact muddy silt	1	6	198	11
Laminated silt with bioturbation	0	1	199	0
Interlaminated dark grey silt and pale grey silty fine-grained sandstone	0	10	199	10
Laminated silt with bioturbation	0	4	200	2
Compact, very dark grey carbonaceous muddy silt	0	11	201	1
Core lost	1	11	203	0
Hard, fine-grained, silty pale grey sandstone, micaceous and ironshot	0	6	203	6
Interlaminated silt and fine sandstone	0	3	203	9
Pale grey compact silt, rapidly darkening and passing down into	1	4	205	1
Olive-grey silty mudstone sporadically ironshot (sphaerosiderite)	3	0	208	1
Olive-grey muddy silt; abundant minute brown sphaerosiderite pellets from 208 ft 6 in to 208 ft 9 in, band of bioturbation, 1 in thick at 210 ft	4	0	212	1
Core lost	0	9	212	10
Interlaminated silt and fine sandstone	0	2	213	0
Compact, dark grey, carbonaceous muddy silt	1	3	214	3
Hard, compact, laminated dark grey silt with thin sandstone intercalations	1	0	215	3
Dark grey, carbonaceous, muddy silt becoming darker and muddy downwards, good plant fragments; finely micaceous	3	2	218	5
Interlaminated silty sand and silt with much plant material	0	11	219	4
Hard, compact, mottled rock; three dimensional spotting of pale sandy material on darker muddy background; no bedding structures; silty at top, becoming more sandy downwards	0	11	220	3
Hard, white, fine-grained sandstone; more silty at top; finely ironshot throughout	1	10	222	1
Core lost	0	11	223	0
Hard, fine-grained, white sandstone	1	7	224	7
As above with intercalations of dark grey muddy silt and bioturbation	1	5	226	0
Compact olive and dark grey silty mudstone; a little plant debris	1	4	227	4
Interlaminated pale micaceous fine sandstone and dark silt; cross-bedding structures with a little bioturbation ..	0	10	228	2
Interlaminated silt and mudstone	1	10	230	0
Interlaminated fine sandstone and silt	1	8	231	8
Dark grey muddy silt with thin paler and coarser intercalations	0	7	232	3
Core lost	0	9	233	0

Description of Strata	Thickness Ft	in	Depth Ft	in
Pellet rock composed of angular pellets, up to 0·2 in across, of greyish green and dark grey siltstone and mudstone set in a matrix of brown ferruginous sandstone	0	2	233	2
Olive-grey muddy silt	0	3	233	5
Olive-grey muddy silt with abundant small brown sphaero-sideritic pellets	0	11	234	4
As above but only poorly ironshot	0	3	234	7
Dark grey muddy silt with lignitic plant fragments and internal slickensiding	0	6	235	1
Hard, compact, olive-grey muddy silt..	1	0	236	1
Very fragmentary sample of olive-grey muddy silt and silty mudstone with blotchy purplish red staining	0	1	236	2
Core lost	6	10	243	0

R.W.G.

Bartley Mill No. 1

Height above O.D. about 211 ft (64·3 m).
Sited 850 yd (777 m) at 196° from Wickhurst [6309 3557]. Drilled in 1958.

Description of Strata	Thickness Ft	in	Depth Ft	in
Head				
Silty and sandy wash containing fragments of weathered silty sandstone .. .. fragments to			3	0
Intimately mixed clayey silt and sandy wash fragments to			8	0
Wealden				
?Upper Tunbridge Wells Sand				
Hard fine-grained silty sandstones .. .. fragments to			12	0
Grey silt and muddy silt with fragments of silty sandstone .. .. fragments to			23	0
? Grinstead Clay				
Finely laminated grey silt and silty shale .. fragments to			26	0
Megaripples of ferruginous medium- to coarse-grained sandstone containing small (up to ¼ in) pellets of siltstone and mudstone	0	1	26	1
Ardingly Sandstone (Lower Tunbridge Wells Sand)				
Dark and light brown fine- to medium-grained sandstone: patchy iron staining	1	0	27	1
Core lost	0	9	27	10
Massive sandstone with thin lenses of coarser sand ..	0	6	28	4
Core lost	8	8	37	0
Massive sandstone	1	8	38	8
Core lost	4	4	43	0
Massive sandstone; micaceous in part; poor plant fragments; patchy iron staining	5	6	48	6
Core lost	3	6	52	0
Fine- to medium-grained massive sandstone: thin grit bands containing small pebbles (up to 3/32 in) at 54 ft 6 in and 55 ft 3 in; dark grey laminated silt with many small rootlets from 54 ft 8 in to 55 ft 0 in; dark brown iron staining from 57 ft 6 in to 58 ft	9	6	61	6
Dark grey massive silty sandstone; poor carbonaceous plant fragments; more silty at base	1	0	62	6

Description of Strata	Thickness Ft	in	Depth Ft	in
Core lost	1	0	63	6
Lower Tunbridge Wells Sand (*undifferentiated*)				
Dark and pale grey laminated compact silt and muddy silt..	1	0	64	6
Fine-grained sandstone; silt and muddy silt, fragmentary sample ..	0	6	65	0
Core lost	6	0	71	0
Dark grey muddy silt weathering to pale grey, abundant plant remains at selected horizons; relatively friable; poorly preserved fine rootlets at 71 ft 9 in: finely micaceous in part; poorly developed sporadic sphaerosiderite spotting from 75 to 76 ft	5	6	76	6
Core lost	10	0	86	6
Olive-grey muddy silt with irregular purple-red staining; passing down into about	2	0	88	6
Compact muddy silt about	1	0	89	6
Core lost	0	6	90	0
Massive, fine-grained, hard white sandstone with darker intercalations of silt; slight iron spotting in part ..	1	0	91	0
Core lost	3	0	94	0
Hard, compact pale grey and greyish green iron-spotted silt and sandy silt ..	1	0	95	0
Massive, iron-spotted silty sandstone	0	6	95	6
Dark grey, compact, muddy silt	1	6	97	0
Core lost	4	0	101	0
Dark grey compact muddy silt	0	6	101	6
Hard, pale greyish green, compact, iron-spotted silt, muddy in part; heavy sphaerosiderite spotting at 105 ft to 105 ft 6 in	4	0	105	6
Interlaminated dark grey muddy silt and silt with fine rootlets and bioturbation	1	6	107	0
Very fine-grained silty pale grey micaceous sandstone with lignitic plant fragments and darker intercalations	2	0	109	0
Pale greyish green silt and muddy silt, hard and compact ..	1	0	110	0
Core lost	2	0	112	0
Thickly bedded, compact, fine-grained sandstone and silty sandstone: minutely ironshot in part, greyish green in colour ..	8	6	120	6
Hard, compact, pale grey and greenish grey muddy silt passing down into very muddy green and khaki silt	11	6	132	0
Interlaminated very muddy dark grey silt and paler greenish silt	2	0	134	0
Fragmentary sandstone sample	1	0	135	0
Core lost	17	0	152	0
Contaminated sample of silt ..	1	0	153	0
Grey muddy compact silt passing down into pale grey micaceous silt	2	6	155	6
Core lost	2	0	157	6

R.W.G.

Bartley Mill No. 2

Height above O.D. about 187 ft (57 m).
Sited 800 yd (732 m) at 194° from Wickhurst [6312 3554]. Drilled in 1958.

Description of Strata	Thickness Ft in		Depth Ft in	

Head

Very sandy honey coloured soil and subsoil.. fragments to			5	0
Fine brown sand .. .. fragments to			8	6

Wealden

?Ardingly Sandstone (Lower Tunbridge Wells Sand)

Fine clean brown sand with fragments of loose sandstone .. .. fragments to			21	6

Lower Tunbridge Wells Sand (undifferentiated)

Muddy grey silt .. .. fragments to			27	0
Pale grey muddy silt	1	0	28	0
Mixed silt and sand with fragments of silt and sandstone .. .. fragments to			30	6
Hard, flaggy green and brown very silty sandstone .. .. fragments to			31	6
Grey silt with enmixed sand .. .. .. fragments to			38	6
Hard, compact, dark grey silt and muddy silt with plant remains .. .. fragments to			44	6
Iron-spotted silt, muddy silt and silty sandstone fragments to			48	0
Iron-spotted silt	2	0	50	0
Core lost	4	6	54	6
Very hard, compact, uniform pale grey silt	4	0	58	6
Core lost	3	6	62	0
Interlaminated fine-grained pale grey silty micaceous sandstone and darker silt; laminations disturbed by bioturbation in part	4	0	66	0
Core lost	2	0	68	0
Hard, compact, dark grey silt	0	6	68	6
Hard, compact silt and muddy silt, paler and more sandy in part	2	6	71	0
Greyish green muddy silt (7 in) passing up into paler grey with greenish blotching, fine-grained, silty sandstone; finely micaceous and ironshot in part	2	6	73	6
Muddy, pale greyish green hard compact silt	1	6	75	0
Silty, fine-grained, hard, micaceous, pale grey sandstone ..	1	6	76	6
Hard, compact, muddy, grey silt	1	9	78	3
Fine-grained silty, pale grey, finely micaceous sandstone with thin muddy intercalations	1	9	80	0
Core lost	2	0	82	0
Hard, compact laminations of pale and dark grey silt and muddy silt	9	0	91	0
Hard, fine-grained, micaceous, white and pale grey, thickly bedded sandstone	2	6	93	6
Very sheared sample; dark olive-green and grey silt, muddy silt and silty mudstone .. .. .. fragments to			98	6
Hard compact dark greyish green silt, partly reddened ..	1	6	100	0
Compact olive sandy silt, micaceous in part and intimately iron spotted	5	0	105	0
Disturbed junction with fine-grained, silty, white, micaceous sandstone	0	6	105	6
Core lost	1	0	106	6
Dark grey muddy silt with more fissile mudstone bands; some disturbed laminae; much plant material at selected horizons, finely micaceous in part and becoming more silty down-				

Description of Strata	Thickness		Depth	
	Ft	in	Ft	in
wards; at 108 ft and 108 ft 6 in beds of medium-grained micaceous sand ½- to 1-in thick occur	2	6	109	0
Hard, compact, pale grey, finely spotted silt; relatively uniform but with thin pale sandy and dark muddy intercalations	6	0	115	0
Core lost	5	0	120	0
Interlaminated pale grey silt and olive greyish green muddy silt; poorly preserved Unionid fragments at 124 ft ..	4	0	124	0
Hard, compact, micaceous, pale grey silt	0	6	124	6
Wadhurst Clay				
Core lost	4	0	128	6
Dark grey shales, mudstones and silty mudstones with thin interlaminated pale grey silts: numerous small rootlets and many vertical stems 0·1 in diameter	1	6	130	0
Dark grey shales as above with some slickensided reddish green mottled mudstone .. .. .. fragments to			145	0
Dark grey shales with numerous thin, hard silt bands .. .. fragments to			150	0

R.W.G.

Appendix II

LIST OF GEOLOGICAL SURVEY
PHOTOGRAPHS

(One-inch Sheet 303)

Taken by Mr. J. Rhodes or Mr. J. M. Pulsford

Copies of these photographs are deposited for reference in the Libraries of the Institute of Geological Sciences, South Kensington, London, S.W.7, and of the Institute of Geological Sciences, Northern England Office, Ring Road Halton, Leeds LS15 8TQ. Black and white prints and lantern slides can be supplied at a fixed tariff, and in addition colour prints and transparencies are available for many of the photographs with numbers higher than 9747.

All numbers belong to Series A

5376	Scenery associated with the outcrop of Hastings Beds in the central Wealden area.
5377	Scenery associated with outcrop of Ashdown Beds near Coleman's Hatch.
5378	Ardingly Sandstone at High Rocks, Tunbridge Wells.
5379	Ardingly Sandstone at High Rocks, Tunbridge Wells.
5380	Jointing in Ardingly Sandstone at High Rocks, Tunbridge Wells.
5381	Ardingly Sandstone on The Common, Tunbridge Wells.
5382	Ardingly Sandstone on The Common, Tunbridge Wells.
5383	Ardingly Sandstone at Denny Bottom near Tunbridge Wells.
5384	The Toad Rock, an earth pillar of Ardingly Sandstone, at Rusthall near Tunbridge Wells.
5885	Ardingly Sandstone at Denny Bottom near Tunbridge Wells.
5886	Honeycomb weathering in the Ardingly Sandstone on The Common, Tunbridge Wells.
6812	Basal beds of the Ardingly Sandstone near Speldhurst.
6820	Uppermost beds of the Ashdown Beds near Cowden.
9748	Scenery associated with the outcrop of the Hastings Beds in the centra Wealden area.
9749	The Toad Rock, an earth pillar of Ardingly Sandstone, at Rusthall near Tunbridge Wells.
9912	Scenery associated with the outcrop of the Ashdown Beds.
10144	Ashdown Beds sandstone as a building stone, Bateman's, Burwash.
10145	Ashdown Beds sandstone as a building stone, Bayham Abbey.
10146	Ashdown Beds sandstone as a building stone, Bayham Abbey.
10272	Basal beds of the Ardingly Sandstone near Speldhurst.
10298	Road cutting in Ardingly Sandstone, East Grinstead.
10299	Lower Tunbridge Wells Sand, Bestbeech.
10300	General view of Ashdown Beds scenery, near Coleman's Hatch.
10301	Gully erosion on Ashdown Beds, near Wych Cross.
10302	Gully erosion on Ashdown Beds, near Wych Cross.

10303   Jointing in Ardingly Sandstone, Eridge Rocks.
10304   Dip slope of the Lower Tunbridge Wells Sand at Blockfield, near East Grin-
        stead.
10305   Ardingly Sandstone near Cowden.
10306   Widened joints in Ardingly Sandstone.
10307   Section in Wadhurst Clay, near Burwash Common.
10308   Valley bulging in Wadhurst Clay.
10312   Junction of the Ashdown Beds and Wadhurst Clay near High Hurstwood.
10313   Gravels of the Third Terrace of the River Medway near Fordcombe.
10314   Gravels of the Third Terrace of the River Medway near Fordcombe.
10315   Section in the Upper Purbeck Beds near Burwash.
10316   Section in the Ashdown Beds near Eridge.
10317   Massive sandrock at the top of the Ashdown Beds near Butcher's Cross.
10318   Festoon bedding in Tunbridge Wells Sand.
10319   Section in the Ashdown Beds near Eridge.
10320   Massive sandrock of the top Ashdown Beds.
10321   Section in the Ashdown Beds near Lamberhurst Quarter.
10322   Scarp of Ardingly Sandstone near Crowborough.
10323   Section in the top Ashdown Beds.
10427   The Toad Rock, an earth pillar of Ardingly Sandstone, at Rusthall, near
        Tunbridge Wells.

INDEX

Printed in England for Her Majesty's Stationery Office by Hull Printers Limited, Willerby, Hull, Yorkshire.
Dd. 502186 K12